Nudibranchs of Southern Africa

A Guide to Opisthobranch Molluscs of Southern Africa

by

Terrence Gosliner

Published by Sea Challengers and Jeff Hamann in association
with the California Academy of Sciences.

SEA CHALLENGERS
Monterey, California

Jeff Hamann
El Cajon, California

E. J. Brill
Leiden, Netherlands

1987

A Sea Challengers & Jeff Hamann Publication

Front Cover: *Nembrotha purpureolineata* by Terrence Gosliner
Technical Editor Dave W. Behrens
Copy Editor Kenneth Hashagen

Printed by DAI NIPPON PRINTING CO., LTD., Tokyo, Japan

ISBN: 0-930118-13-8 (Sea Challengers) ISBN: 90-04-08420-7 (E.J. Brill)

Library of Congress Catalog Card Number: 86-062429

First Edition

SEA CHALLENGERS
4 Somerset Rise
Monterey CA 93940

JEFF HAMANN
475 West Bralley Ave.
El Cajon CA 92020

E. J. BRILL
Plantijnstraat 2
Leiden Netherlands

Phototypesetting and pre-press production by Padre Productions, San Luis Obispo CA.

Foreword

To the professional biologist, the opisthobranchs represent an interesting evotionary offshoot of the Mollusca, in which the precarious balance between the hard and the soft body parts has become heavily tilted in favor of the latter. The earliest opisthobranchs were probably all burrowing infaunal animals and certainly the most primitive living opisthobranchs are burrowers. The head became a flattened spade while the tentacles were lost or shifted to the sides or rear of the head. The shell became streamlined in shape, the operculum was usually lost, and great lateral (parapodial) lobes of the foot gradually developed, helping to enhance streamlining and to exclude foreign particles from the shell and mantle cavities.

From these early burrowing opisthobranchs with their increasing emphasis upon chemical and biological defense, and their concomitant reduction in the importance of the shell and operculum, arose a number of distinct lines of descent, the modern representatives of which constitute several groups of species, genera, and families which have taken up new and quite different dietary specializations. At the present time, the majority of opisthobranch species live on the sea bottom, not within it or above it. They are the most visually exciting invertebrate animals, with an unbelievable flamboyance in their body ornamentation. They have agile, soft bodies, perfectly adapted chemosensory tentacles and buccal apparatus, and virtual immunity to attack by fish and many other predators, while the searching ciliated larvae have an uncanny ability to recognize and then metamorphose upon the adult diet.

The South African coasts have everything to offer the student of mollusks. Opisthobranch specialists, especially, have been attracted to the area throughout the century, starting with the pioneering studies of Bergh (1907) and O'Donoghue (1929), who made their investigations under Europe's grey skies using specimens shipped to them by South African naturalists. The great South African malacologists, Barnard (1927) and Macnae (who published a series of scientific papers during the fifties and sixties), were brilliant field biologists as well as perceptive and meticulous laboratory workers. And now Dr. Terrence Gosliner has addressed his attention to this fauna, opened up for comprehension by the earlier studies, and now ripe for development and review by this distinguished international authority, who has a working knowledge of opisthobranchs from the Pacific, Atlantic, and Indian oceans. This book stands as a landmark in the progress of the subject, with its magnificent color photographs from life and its completely authoritative text.

T.E. THOMPSON

Ph.D., D.Sc., Reader in Zoology in the University of Bristol.

Dedication

This book is dedicated to Eveline Marcus for her monumental contributions to the knowledge of opisthobranchs and for her friendship and generosity.

Contents

Foreword . . . iii
Acknowledgments . . . vi
Photo Credits . . . vi
Introduction . . . 7
 Evolutionary History . . . 7
 Defense Mechanisms . . . 7
 Feeding . . . 9
 Reproduction . . . 9
 Systematics . . . 11
 Collection of Opisthobranchs . . . 12
 Biogeography . . . 13
Identification of Opisthobranchs . . . 16
 The Species of Southern African Opisthobranchia . . . 16
 Glossary . . . 21
 Key to the Opisthobranchs of Southern Africa . . . 29
Species Accounts . . . 39
References . . . 130

Acknowledgments

Publication of this book has been made possible by the assistance of numerous people. Bill Liltved has contributed significantly to every phase of the production of this work. His keen observational abilities in the field have provided much of the South African opisthobranch fauna. He also drew all of the illustrations. His help and enthusiasm are greatly appreciated.

Publication of this book would not have been possible without the financial support and enthusiasm of Jeff Hamann. The continued interest and moral support of Dave Behrens made this book a reality. His help and outstanding editorial comments are greatly appreciated.

My wife, Bonnie, has also provided tremendous assistance in the collection of animals and the typing and proof reading of the various drafts of the manuscript. Michael Gosliner devoted many hours to proof reading the text. Patricia Dal Porto typed the final draft of the book. Their help and encouragement have been instrumental in the successful completion of this work.

Many other individuals have given generously of their time and have aided in the collection of specimens including Gary Williams, Clarice Connolly, Robbie and Charles Griffiths, Gail and Mike Strong, Michelle and Schalk van der Merwe, Brian Hayes, Anna Leakey, Graham Avery, Cathy and George Car, Charles Hirons, Michael Bougaardt, Bettie Louw, Butch Hulley, and Sidney Kannemeyer.

This work was undertaken at the South African Museum and many staff members aided in the production of this book. The late Thomas H. Barry, former Director, provided enthusiasm and interest which greatly facilitated the production of this work.

The Natal Anti-shark Measures Board has generously provided support for field work on the Natal coast. Beulah Davis, Graham Charter and Theo Meyer have assisted greatly.

Much of the field work was supported by grants from the Council for Scientific and Industrial Research.

To these individuals and institutions, as well as others I have undoubtedly overlooked, I express my sincere appreciation.

Photo Credits

All photographs are by the author except:

Species #	*Photographer*
103, 177	Gary Williams
138, 156	Bill Liltved
240	Charles Griffiths

Introduction

Evolutionary History

Opisthobranch snails are among the most strikingly beautiful animals that inhabit the world's oceans. Their vivid colors, arranged in a myriad of patterns, defy the imagination. A great diversity of opisthobranchs inhabit all marine environments around southern Africa, from the warm tropical reefs of Natal to the temperate kelp forests of the Atlantic coast of the Cape Province.

Unlike the more typical marine snails, (prosobranchs), which are popular among shell collectors, opisthobranchs exhibit an evolutionary trend towards the reduction, internalization, and complete loss of the shell. The shell has taken different forms in various evolutionary lines of opisthobranchs (Figure 1). It is apparent that the reduction and loss of the shell have occurred independently in several different groups of opisthobranchs. Opisthobranchs exhibit several other trends that distinguish them from their prosobranch ancestors. Durings its development, the body of a prosobranch undergoes a radical transformation called torsion. The viscera, along with a gill cavity, rotate 180° from their initial position so that the gill and anus are directed towards the front of the animal. Opisthobranchs exhibit a reversal of this process (detorsion) to varying degrees. The net result is that the gill cavity and anus are situated on the right side or posterior end of the animal. Most prosobranchs have separate sexes, while most opisthobranchs are hermaphroditic, possessing both female and male reproductive organs in the same individual. Although opisthobranchs are hermaphroditic, they are incapable of self-fertilization. The major sensory organ of prosobranchs is a highly folded structure called the osphradium, situated in the gill cavity. Water currents pass over the osphradium, enabling the snail to determine aspects of the chemical composition of its environment. Only the most primitive opisthobranchs retain an osphradium and the vast majority of opisthobranchs have developed other sensory structures, such as sensory mounds or oral tentacles, around the mouth and paired, horn-like rhinophores on the top of the head. The advantage of these structures to opisthobranchs is that they are more closely associated with the head region and therefore the brain. Generally speaking, opisthobranchs have more nervous tissue concentrated in the head, thus a proportionately larger brain.

Defense Mechanisms

The molluscan shell evolved to protect the soft bodied animal from being eaten by predators. Typical prosobranch snails utilize the shell in this manner and can retreat into their calcified fortresses, which can be sealed by a trap-door, the operculum. A few primitive opisthobranchs retain a well developed shell and operculum, but the vast majority of forms exhibit varying degrees of modification and loss of the shell. Along with the reduction or loss of the shell and its protective value, opisthobranchs have evolved other defense mechanisms. Most opisthobranchs defend themselves by the secretion of toxic chemicals like sulfuric acid or by firing stinging cells derived from other animals on which they may feed.

Coloration in opisthobranchs, like that of butterflies, birds, and coral reef fishes, is not just for our aesthetic enjoyment, but has an important biological function. Brightly colored animals with distinctive patterns usually exhibit warning coloration to potential predators. Warning, or aposomatic, coloration requires the predator to experience the unpleasant taste of the prey and to associate the specific color pattern with an unpleasant encounter. Fish are probably the only potential predators that have sufficient intelligence and visual acuity to recognize and avoid the specific color patterns of opisthobranchs. When a nudibranch floats in the water, grazing fish such as hottentots (*Pachymetopon blochii*) usually rapidly engulf it and even more rapidly spit it back out, generally in an unharmed condition. Many opisthobranchs are so brilliantly colored and extroverted in their behavior, often crawling on a rock in the open, that there is little question they are employing warning coloration, advertising their obnoxiousness. However, this does not always seem to be the case. Other species are extremely difficult to see and blend in with their natural surroundings, a defense mechanism termed cryptic coloration. *Doris verrucosa* is the same color as the bread crumb sponge on which it feeds and it is often embedded in small depressions in the sponge colony. *Corambe* sp. feeds on bryozoan colonies; the striations on its back closely match the lace work pattern of the bryozoan. Similarly, the eggs of this nudibranch blend in with the surrounding colonial network. Whether a species is exhibiting warning coloration or cryptic coloration may depend largely on the substrate where it is situated. *Phestilla melanobranchia* is a species which may be black, yellow, or orange depending on the species of coral on which it feeds. If the nudibranch happens to be on its coral prey, it is well camouflaged. If it is crawling out in the open, perhaps in search of a mate, it is often very obvious and would then be exhibiting warning coloration. Many opisthobranchs also avoid predation by swimming. *Hexabranchus sanguineus*, a vivid red dorid, swims by undulating its body from front to rear. *Bornella anguilla* undulates its body from side to side. Swimming can be sustained for several minutes in these species. This method of escape would be particularly effective against crustaceans, which usually remain on the bottom rather than in the water column. The more alternatives a species has in defending itself, the better adapted it is in avoiding predation and, therefore, has a better chance of survival. While coloration in opisthobranchs is one of their most outstanding features, it is also an adaptation that is critical to their survival.

Figure 1. Reduction of shell in various opisthobranch taxa.

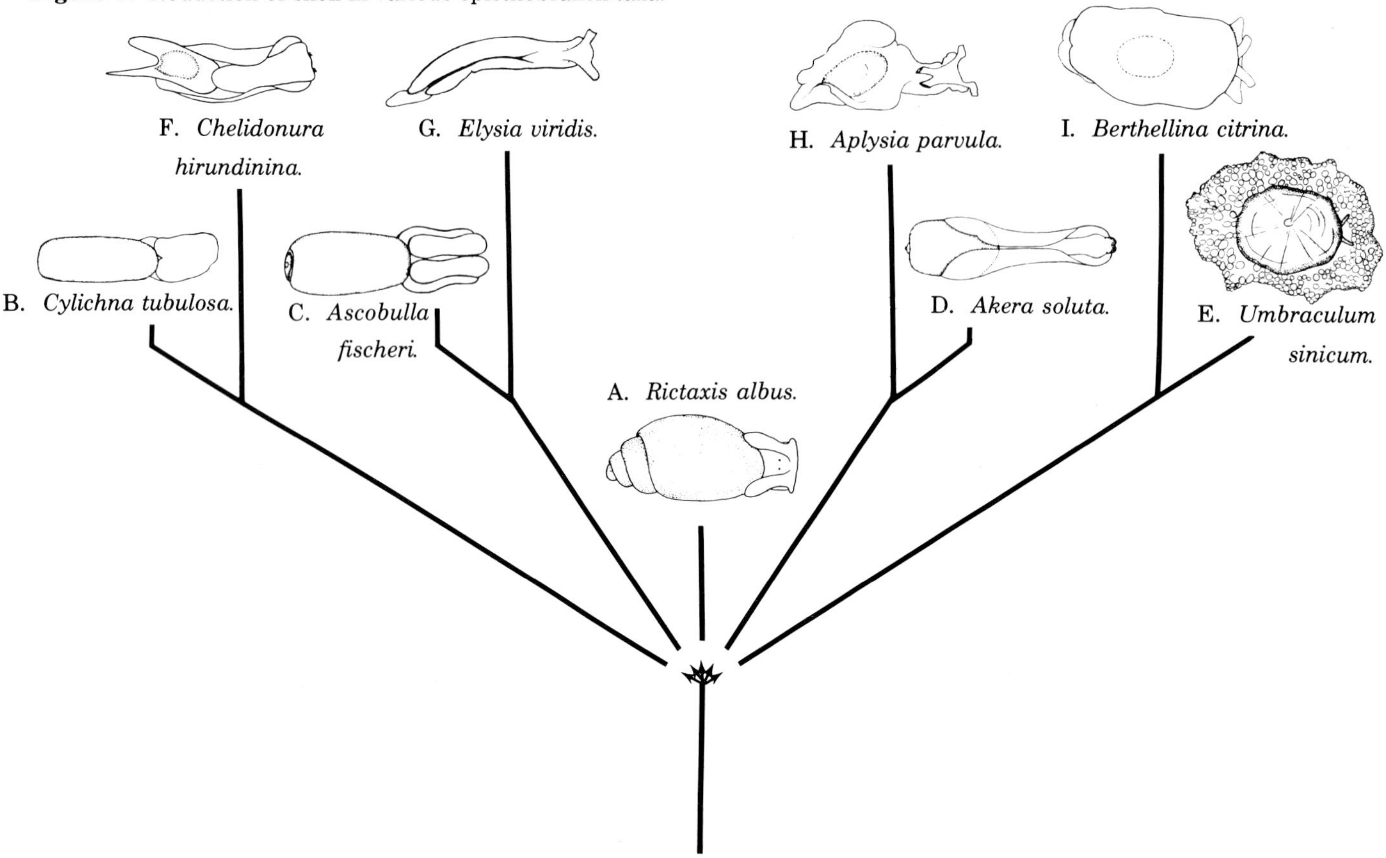

Figure 2. Radular teeth of various opisthobranchs.

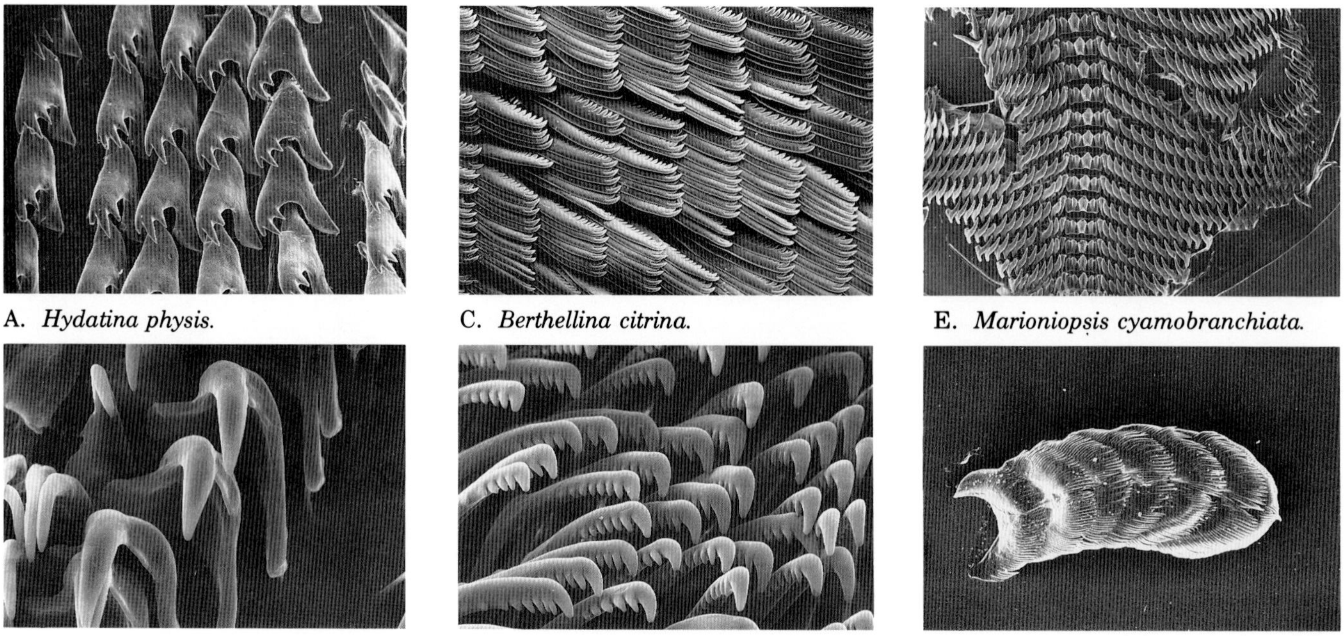

A. *Hydatina physis.*

B. *Gastropteron flavobrunneum.*

C. *Berthellina citrina.*

D. *Chromodoris quadricolor.*

E. *Marioniopsis cyamobranchiata.*

F. *Baeolidia palythoae.*

Feeding

Opisthobranch mollusks feed on a variety of organisms rarely fed upon by other animals. Some species, such as *Haminoea alfredensis* and *Philine aperta*, are generalized grazers on algae or microscopic animals. The majority of forms are highly specialized herbivores or carnivores and many restrict their diet to a single genus or species of plant or animals. The variety of organisms fed on by opisthobranchs is indicated in Table 1. The vast majority of opisthobranchs feed using a pair of jaws and a ribbon of teeth called a radula. The jaws and radular teeth are made of chitin, the same hardened protein which forms the carapace of crustaceans such as crabs, lobsters, and crayfish. The configuration and structure of the radular teeth in opisthobranchs are highly variable (Figure 2). Generally, the radula is broad in grazing herbivores such as the sea hare, *Aplysia*, and dorid nudibranchs like *Chromodoris*, which rasp tissue from sponges. The radula is somewhat narrow in *Notobryon wardi*, which eats hydroids. In *Elysia* and other sacoglossans there is only a single tooth per row of radular teeth. The sacoglossans puncture individual algal cells with the sharp teeth and suck out the liquid contents. To supplement their diet sacoglossans usually retain chloroplasts, the intracellular organelles which are responsible for photosynthesis in plant cells. These chloroplasts remain active in the translucent skin of the digestive branches and continue to produce sugars which are consumed by the sacoglossans. *Placobranchus ocellatus*, which is found throughout the Indian and Pacific oceans, can survive for months without any food other than the sugars produced by the chloroplasts it retains beneath its skin. *Baeolidia palythoae* and most aeolid nudibranchs also possess only a single tooth per radular row. In this case the comb-like teeth are used to scrape tissue from zoanthid sea anemones. The variability of radular form in opisthobranchs is far greater than that of their prosobranch cousins and attests to their many specialized modes of feeding.

Not all opisthobranchs have a radula. Species of *Philinopsis* feed on other opisthobranchs by engulfing their prey whole with their highly muscular pharynx. This is accomplished by rapid suction like a vacuum cleaner. *Melibe rosea* also lacks a radula but has a tentacular hood around the mouth, which it uses to trap small crustaceans. Dorids of the genera *Dendro*-doris, *Doriopsilla*, and *Phyllidia* also lack a radula and suck tissue of sponges.

Reproduction

Although reproduction in opisthobranchs requires copulation between two individuals, fertilization may not take place immediately. A specialized organ called the seminal receptacle can store sperm from another individual for several months until the eggs have reached maturity. When the eggs are mature they pass by the seminal receptacle and are fertilized. They subsequently are surrounded by albumen, which provides nutrients to the developing embryo. Next, the eggs are encapsulated in a thin, transparent membrane. Generally, there is

Table 1. Specialized Diets of Some Southern African Opisthobranchs

Prey	Opisthobranch predator
Algae	
Codium spp.	*Elysia viridis, Placida dendritica*
Caulerpa spp.	*Oxynoe viridis, Lobiger souverbiei*
Protozoa	
Forminifera	*Retusa truncatula*
Porifera (sponges)	
Hymeniacedon perlevis	*Doris verrucosa*
Haliclona sp.	*Hypselodoris capensis*
Cnidaria	
Hydrozoa (hydroids)	
Tubularia spp.	*Catriona casha, Facelina olivacea*
Sertularella sp.	*Cuthona speciosa*
Eudendrium sp.	*Flabellina funeka, F. capensis*
Chondrophora	*Glaucus atlanticus*
Velella sp.	" "
Porpita sp.	" "
Siphonophora	
Physalia sp.	" "
Anthozoa	
Actinaria (sea anemones)	
Anthothoe stimpsoni	*Aeolidiella indica*
Zoantharia (zoanthids)	
Palythoa spp.	*Baeolidia palythoae*
Gorgonacea (sea fans)	
Eunicella verrucosa	*Tritonia nilsodhneri*
Melitodes sp.	*Phyllodesmium serratum*
Alcyonacea (soft corals)	
Xenia sp.	*Marioniopsis cyanobranchiata*
Sarcophytum trochiforme	*Dermatobranchus* sp. 1
Madreporia (true corals)	
Tubastrea aurea	*Phestilla melanobrachia*
Platyhelminthes (flat worms)	*Chelidonura hirundinina*
Annelida (segmented worms)	
Cirratulidae	*Hydatina physis, H. amplustre*
Mollusca	*Philinopsis cyanea,*
other opisthobranch species	*Pleurobranchaea tarda*
opisthobranch eggs	*Favorinus japonicus, F. ghanensis*
Arthropoda	
Cirripidea (barnacles)	
Lepas sp.	*Fiona pinnata*
Amphipoda	*Melibe rosea, M.* sp.
Bryozoa (moss animals)	
Bugula neritina	*Polycera hedgpethi, P. capensis*
Membranipora spp.	*Limacia clavigera, Corambe* sp.
Chordata	
Ascidiacea (tunicates)	*Goniodoris mercurialis, G. castanea*

Figure 3. Egg masses of various opisthobranchs.

A. *Doris verrucosa.*

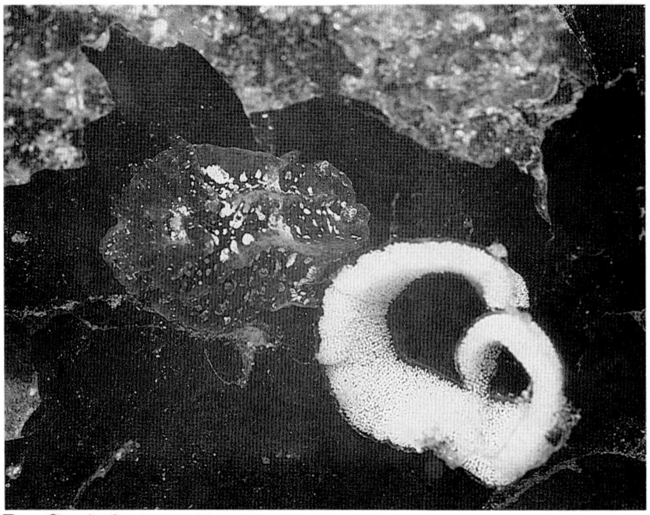

B. *Goniodoris castanea.*

C. *Pleurobranchaea bubala.*

D. *Limacia clavigera.*

E. *Doto pinnatifida.*

a single egg per capsule, but in some instances, such as in *Janolus capensis*, there may be as many as 45 eggs per capsule. Immediately prior to the emergence of the egg mass from the female reproductive opening, a thick layer of mucus is added; the mucus surrounds the capsules and links them into a single matrix. The egg mass is then deposited on the substrate. The shape of opisthobranch egg masses is exceedingly variable (Figure 3) and can often be used to distinguish closely related species. Within the egg mass, the embryo develops for a period of a few days or as long as several weeks before hatching into a larva.

Opisthobranch mollusks are slow moving creatures and can travel only limited distances during their adult life. Their dispersal to other suitable environments is dependent upon a planktonic larval stage called a veliger. Opisthobranch veligers have a thin oval or oblong shell even in those forms which lack a shell as an adult. Veligers may extend their life in the plankton if no suitable substrate for their adult life is present. Usually a specific environmental cue triggers metamorphosis. The larvae of *Phestilla lugubris*, which feeds on corals, must come in contact with coral mucus to initiate the change from a swimming larva to a crawling juvenile. This flexibility in the length of larval life increases the chances that larvae will successfully settle and permits them to cross entire ocean basins. Some exceptional species of opisthobranchs (only *Embletonia gracilis* in southern Africa) have direct development in which the egg develops into a juvenile, bypassing the free-swimming veliger stage.

Systematics

The subclass Opisthobranchia is divided into several major orders based on the external and internal morphology of the animal.

Order Cephalaspidea (=Bullomorpha)

The opisthobranchs that most closely resemble their prosobranch ancestors belong to this order. The vast majority of cephalaspideans possess a shell, although it may be reduced and internal. The name cephalaspidean means head-shield and all members of this group have a well-developed head-shield which is used to plow beneath the surface of the sand. The alternate name for the order, Bullomorpha, is misleading, as not all cephalaspideans have a bulloid shell. Simiarly, members of some other orders such as *Akera* (Order Anaspidea) and *Cylindrobulla* (Order Sacoglossa) retain a bulloid shell. Members of the Cephalaspidea are the most morphologically diverse of all opisthobranchs. The least modified forms, like *Japonacteon* sp., have a shell with an elevated spire. Species such as *Hydatina amplustre*, *Cylichna tubulosa*, and *Haminoea alfredensis* have an external shell with a bulloid spire. *Philine aperta* and members of the Aglajidae, including *Chelidonura fulvipunctata*, have a reduced internal shell. A shell is absent in some members of the Runcinidae and Gastropteridae.

In addition to the variability of the shell that occurs in the Cephalaspidea, the gill or mantle cavity may be situated in several different positions. In *Japonacteon* it is directed anteriorly. In most other cephalaspideans the mantle cavity faces the right side of the animal. In *Chelidonura* and *Philine* it is directed posteriorly. These are changes that are associated with the process of detorsion. Aspects of the internal morphology of the digestive, nervous, and reproductive systems exhibit similar trends of modification.

The vast majority of Cephalaspidea live just beneath the surface of the sand and are rather dull in their coloration. However, *Micromelo*, *Hydatina* and *Chelidonura* are often brightly colored and are seen crawling on the surface of sand or rock. These genera prey on other large invertebrates and have well-developed sensory structures to detect prey. *Chelidonura*, in addition to having well developed eyes on the anterior end of the head, has bundles of sensory cilia around its mouth. With these cilia it is able to track its prey by following its victim's mucus trail.

Order Anaspidea (=Aplysiomorpha)

Members of the Anaspidea are commonly known as sea hares because of the rabbit-like appearance of some members of this order. The most primitive anaspidean, *Akera*, closely resembles a "bulloid" cephalaspidean but has several unique, derived features in common with the sea hares. For this reason it is placed in the Anaspidea following Ghiselin (1966) and Gosliner (1981), although not all authorities are in agreement about its systematic position. Other members of the order generally also have large parapodia that are folded over the mantle cavity located on the back of the animal. *Akera* retains a bulloid shell, whereas all other anaspideans have a small internal shell plate or no shell at all. Most sea hares defend themselves by secreting a purplish ink when disturbed. Anaspideans are exclusively herbivorous. *Aplysia* feeds on large algae while *Bursatella* ingests large amounts of sand, much as an earthworm does. Despite their bulky appearance, several species of anaspideans are capable of swimming, including two South African species, *Akera soluta* and *Aplysia oculifera*.

Order Sacoglossa (=Ascoglossa)

All but three sacoglossans are herbivorous. The only exceptions are one species from North America *(Olea hansineensis)* and two from Europe (*Calliopaea bellula* and *C. oophaga*) that feed on the eggs of other opisthobranchs. Sacoglossans are variable in form and may have a bulloid external shell, as in *Ascobulla* and *Volvatella*, or may entirely lack a shell as an adult, as in *Elysia* and Placida. Some of the most unusual sacoglossans are the "bivalved gastropods" such as *Berthelinia schlumbergeri* and *Julia zebra*. It was once thought that these opisthobranchs might represent a "missing link" between the gastropods and bivalves. However, it has since been demonstrated that the larval stage has a typically coiled gastropod shell and that the bivalved condition has evolved secondarily. Most sacoglossan groups are associated with green algae of the genus *Caulerpa*, suggesting that most of the evolution and radiation of the order occurred in association with *Caulerpa*. Sacoglossans are characterized by a radula with a single tooth per row which is used for piercing algal cells and sucking out their liquid contents.

Order Notaspidea (=Pleurobranchomorpha)

All notaspideans have a gill situated on the right side of the body. In the more primitive forms such as *Umbraculum sinicum*, there is a thickly calcified, plate-like shell. In the majority of species the shell is a reduced internal plate.

Pleurobranchaea is one of the most highly modified notaspideans and it lacks any vestige of a shell as an adult. Notaspideans are exclusively carnivorous and utilize strong jaws and a broad radula to scrape tissue from their sponge or tunicate prey. *Pleurobranchaea* is a voracious predator and feeds on a variety of large prey including other opisthobranchs and fish. Notaspideans usually secrete sulfuric acid as a defense mechanism. Most authorities agree that notaspidean ancestors gave rise to the largest order of opisthobranchs, the Nudibranchia.

Order Nudibranchia

The Nudibranchia constitute the overwhelming majority of opisthobranchs and in South Africa represent three quarters of the total number of opisthobranch species. All nudibranchs lack shells as adults. They exhibit an incredible range of external form. Nudibranchs can be subdivided into four sub-orders.

Suborder Doridacea

There are more species of dorids than in all the other suborders of nudibranchs combined. Most species are distinguished by a circle of gills that surrounds the anus on the posterior end of the back (notum). Three groups of dorids lack a circle of gills. *Corambe* has its gills situated in a posterior notch between the notum and foot. Members of the Phyllidiidae lack gills around the anus and have a series of gill-like folds along the sides, between the notum and foot. *Okadaia* entirely lacks gills. Dorids contain three major groups. In the cryptobranchs and dendrodorids the body is oval and the gills can be retracted into a distinct sheath. Cryptobranchs have a well-developed head and scrape tissue from sponges using a broad radula. Porostomes have a poorly developed head without tentacles; they lack a radula but also feed on sponges. The third group, the phanerobranchs, have an elongate body and there is no sheath into which the gills can be retracted. They have a narrower radula than cryptobranchs and generally feed on bryozoans and tunicates.

Suborder Dendronotacea

Members of this suborder are characterized by rhinophores with a cup-like sheath. They also have a row of gills or cerata-like processes along the margins of the notum. Dendronotaceans feed exclusively upon cnidarians, with the exception of *Melibe*, which is a filter feeder. The Tritoniidae, which have tufts of gills along their notal margin, feed exclusively on alcyonaceans such as gorgonians and soft corals. The Scyllaeidae, Dotoidae, and Bornellidae are specialized predators of hydroids. All dendronotaceans except *Doto* have well-developed jaws. *Melibe* and *Fimbria* are the only dendronotaceans that lack a radula.

Suborder Arminacea

The Arminacea are the smallest suborder of nudibranchs, although they are well represented in the waters of southern Africa. They are generally characterized by an expanded oral veil or velum on the head. In the Arminidae the notum contains longitudinal striations. Gills may be present or absent. If present they are found along the sides of the animal between the notum and foot. Arminids feed on soft corals and pennatulaceans (sea pens). The Janolidae possess well-developed cerata and feed exclusively on bryozoans. Members of the Lemindidae have well-developed parapodia and lack gills or cerata. They appear to be restricted to southern Africa and little is known about their biology.

Suborder Aeolidacea

The aeolids are the second largest nudibranch suborder. They all have well-developed cerata and most species store nematocysts in the ceratal tips. These stinging cells are collected from their cnidarian prey in an undischarged state and can be utilized by the aeolids for their defense. The vast majority of aeolids like the sacoglossans have a single radular tooth per row. *Flabellina* and *Eubranchus* retain a pair of lateral teeth on either side of the central, or rachidian, tooth. Aeolids also have a pair of strong chitinous jaws. While some aeolids prey upon other opisthobranchs or their eggs, bryozoans, sea anemones, corals, or gorgonians, the vast majority of species feed on hydroids. Some aeolids may restrict their diet to a single species of hydroid, while others, like *Cratena capensis*, will feed on virtually any hydroid.

Collection of Opisthobranchs

Opisthobranchs can be found in most intertidal and shallow water habitats all along the coast of southern Africa (Map), with the exception of wave swept, sandy beaches. Seeking out the specialized habitats of individual species is the challenge that faces anyone interested in studying opisthobranchs. Different methods must be employed in different environments.

In order to find small cephalaspideans it is necessary to sift fine sand with a sieve. Patience and a keen eye are necessary to find some species such as *Retusa truncatula*, which reaches a maximum length of 4 mm. Larger cephalaspideans and *Akera soluta* are most commonly found in estuaries or tidal swimming pools. They are most easily found by following their burrow trails through the sand and digging or fanning the sand at the end of the trail. Egg masses and empty cephalaspidean shells on the surface of the sand are good indicators that these animals are present. In estuaries, eel grass beds and clumps of algae often yield a variety of opisthobranchs, especially sea hares.

Most opisthobranchs are inhabitants of rocky environments rather than sandy ones, since their prey requires a permanent and stable substrate on which to grow. During spring low tides a great diversity of opisthobranchs can be found in the lower fringes of the intertidal zone. The best place to look is underneath large rocks, among sponges, bryozoans, tunicates, and hydroids. When turning rocks one must remember to carefully return them to their original position. It is no accident that the organisms living under rocks are different from the ones on the surface. It is the only place where they can survive. If left exposed they will not be there for future study and observation. When looking on the under surface of rocks it may not always be obvious that nudibranchs are present; they may be cryptic in their appearance. Again, look for egg masses, one of the best indicators that nudibranchs can be found nearby. In the intertidal, a variety of sacoglossans can be found by examining green algae. *Caulerpa* and *Codium* are usually the best algae to examine. They must be closely scrutinized as most sacoglossans are the same color as the algae they inhabit.

It is often productive to place a bit of *Caulerpa* or *Codium* in a bucket of sea water for a few hours. The sacoglossans will generally start crawling into the open after a period of time. The same procedure can be followed with hydroids or bryozoans. This often yields a variety of small nudibranchs which may have been overlooked in the field.

The vast majority of opisthobranch species are found just beyond the lower fringes of the intertidal, in the shallow subtidal. The way to observe the greatest number of species is by scuba diving or snorkeling. The colonial animals on which nudibranchs feed are most abundant on vertical walls and in caves. Most species of nudibranchs are brightly colored and can be readily observed. Finding smaller or more cryptic species requires a vigilant eye, which comes only with experience and a knowledge of what particular opisthobranch species might be found on a specific prey species. On a single dive the inexperienced observer should be able to find 10 to 15 species of opisthobranchs without too much difficulty.

People are often curious about the best way to preserve the color of opisthobranchs after they have been collected. The answer is very simple. Photography or drawing are the only reliable ways to capture the grace and beauty of opisthobranchs. An opisthobranch specimen in a bottle of alcohol or formalin is usually a greyish, contracted blob which is of interest only to the serious researcher. The best way to study opisthobranchs is to observe them while they are alive. Small species must be observed with a strong magnifying glass or, ideally, with a dissecting microscope. Following their observation, common species should be returned to the habitat from which they were collected.

Specimens of rare or undescribed species are always needed and appreciated by scientists. In this case the specimens must be photographed and preserved. The best way to preserve opisthobranchs it to narcotize them prior to killing them as they do not contract as much. This can be accomplished by freezing them or by gradually adding a solution of 8% magnesium chloride to a small amount of sea water containing the specimen. The latter process is more time consuming but produces better results. After the specimen is narcotized and does not contract or move when probed, it is ready for preservation. The best preservative is Bouin's fixative or, alternatively, a 10% solution of formalin. Barring this, 70% ethyl alcohol, which is readily obtainable, can be used,but is less desirable as it does not preserve the tissue as well.

Biogeography

As most opisthobranchs have a planktonic veliger larva which can delay metamorphosis until the adult food source is present, they can be transported long distances by oceanic currents. This dispersal ability accounts for the fact that many species have very wide distributional ranges.

Any study of the distribution of a group of organisms is limited not only by the amount of collecting which has been done within the particular geographical area of interest, but also by the extent of knowledge of the faunas in surrounding oceans. There are vast areas of the world where the opisthobranch faunas are virtually unknown. Owing to this fact, it is possible that our view of their biogeographical relationships may change as we learn more about these regions.

In studying the biogeography of an area it is not adequate to simply know what species are present or absent in that region. One must know what ecological role a species plays within a locality. Some idea of its abundance and prey/predatory relationships help ascertain the impact of a single species in a particular area. One must also know whether a species is present in an environment throughout the year or whether it only sporadically or seasonally influences that habitat. It is important to know if a species reproduces in a given habitat, as this drastically affects its impact over a prolonged period of time. Like any study of living organisms in their natural environment, biogeographical data must be painstakingly accumulated over many years.

Another factor that must be considered when studying the biogeography of organisms is whether a species' presence in an area is a result of natural dispersal or whether man's influence has altered its distributional pattern. For example, the aeolid nudibranch, *Catriona columbiana*, has been found in southern Africa only from the pilings in Cape Town Harbor. It has been recorded from the Pacific coast of North America, Japan, and possibly from Auckland Harbor in New Zealand. The prey of nudibranchs are frequently colonial organisms that foul ship bottoms. Nudibranchs (often with their food and eggs) can be transported long distances on the hulls of ships. This appears to be the manner in which *Catriona columbiana* was introduced into South African waters. Two other cases of unusual distributional patterns do not offer such simple explanations.

Thecacera pennigera was originally described from England but has been subsequently found near harbors in Brazil, Japan, Australia, and New Zealand. Willan (1976) suggested that the species owes its distribution to human intervention through shipping. Barnard (1933) described *T. pennigera* (as *T. lamellata*) from a single specimen from Cape Town Harbor. However, *T. pennigera* is commonly found along the coast of southern Africa from Cape Town to Umgazana, Transkei. As several localities, including Umgazana, are over 250 km from the nearest harbor, it is difficult to attribute the distribution of this species in southern Africa solely to introduction by shipping.

Polycera hedgpethi has been recently reported from southern Africa (Gosliner 1982) and Australia and New Zealand (Willan and Coleman 1984) although it was previously known only from the Pacific coast of North America. South African specimens were collected in the Keurbooms River Estuary near Plettenberg Bay. This estuary is shallow and certainly is not subject to international shipping. Therefore, it is unlikely that this species was introduced to South Africa, as suggested by Willan and Coleman.

One species of South African nudibranch appears to have been introduced to other parts of the world. *Polycera capensis*, which is very common along the coast of southern Africa, is also known from Australia, where it is restricted to the vicinity of Sydney Harbor. Specimens have often been collected from the hulls of ships. Thompson (1975) noted that it was strange that the species had not been collected from Australia until 1932, although Angas had collected extensively in the area in the mid-19th century. It is highly probable that the species was introduced to Australia from South Africa by ships after Angas had conducted his survey of the nudibranch fauna.

The overwhelming majority of opisthobranch species have been distributed by natural means. Along the coast of southern Africa it is readily apparent that there are two distinct opisthobranch faunas. A tropical fauna is found from Mozambique, extending southwards through Natal and Transkei, terminating somewhere near Port Alfred. The southern boundary of this fauna is not precise and fluctuates seasonally. Occasionally, warm tropical waters extend as far south as Algoa Bay. When this occurs, tropical opisthobranchs are occasionally found near Port Elizabeth or even as far south as Knysna. As different opisthobranchs have different temperature limits, there appears to be a gradual dropping-out of species as one moves southward. Shallow bays, such as Algoa Bay and Knysna Lagoon, which have periodic influxes of warm water, serve as isolated refuges for tropical species near their southern limit.

The second major faunal province extends along most of the coast of the Cape Province from Port Alfred around Cape Point and extending at least as far north as the Namibian border. Few intertidal specimens, and virtually no shallow subtidal specimens, have been collected along the Atlantic coast of the Cape Province north of Eland's Bay. Within this vast geographical area species are limited to varying portions of the province. *Melibe rosea* and *Notobryon wardi* are found throughout the 2000-km coastline of the Cape Province. *Dendrodoris caesia* and *Volvatella laguncula* are apparently limited to 800 km of coast between False Bay and Port Elizabeth. Many people have sub-divided the Cape Faunal Province into two distinct provinces, with a sharp boundary at the tip of the Cape Peninsula. Cape Point does not appear to form a sharp faunal boundary in the case of opisthobranchs. Of the species found on the Atlantic coast of the Cape Peninsula, 73% are also found in False Bay.

The biogeographical affinities of the opisthobranch faunas of southern Africa can be divided into four major categories. The first is the endemic fauna. Endemic species are those which appear to be restricted to a particular geographical area. In some cases, species appear to be endemic to an area solely because they have not been found in adjacent regions that are poorly studied. Only through prolonged and thorough study of an area can the endemic component of a fauna be determined with reasonable certainty. The second significant faunal element is the circumtropical component. This group consists of species which are distributed throughout the tropical portions of the Atlantic, Pacific, and Indian oceans. Thirdly, is the Indo-Pacific faunal component, consisting of species which are widely distributed in the tropics of the Indian and Pacific oceans but are absent from the Atlantic. Lastly, are those species which are known only from southern Africa and other portions of the Atlantic Ocean. Most species of the Atlantic component are also found from European and Mediterranean waters.

If one examines the biogeographical affinities of seven areas along the coast of southern Africa (the Atlantic coast of the Cape Peninsula, False Bay, the Knysna vicinity, Port Elizabeth, Transkei, Durban, and the Sodwana Bay region) striking differences can readily be observed (Table 2). Most significant are the differences between the relative proportion of the faunas that endemic and Indo-Pacific. Two-thirds of the fauna of the Atlantic and False Bay are endemic to southern Africa, while 1 to 6% of the species are found in the Indo-Pacific. This is consistent with other areas of the world. Most temperate faunas, such as that found in the Cape Province, are highly endemic, while tropical faunas have a very small endemic component.

It is readily apparent that the tropical fauna of Mozambique, Natal, and Transkei originated from the deep tropics of the Indo-Pacific, but the origins of the largely endemic Cape fauna are not so obvious. Looking at the various genera of endemic Cape species does provide some clues. Genera such as *Cadlina*, *Aldisa*, *Rostanga* and *Eubranchus* are largely temperate, with representatives in both the northern and southern hemispheres; until recently, *Aldisa* has been known only from the northern hemisphere. On the other hand, *Gargamella* and all but one species of the genus *Aphelodoris* are restricted to the Subantartic and cold temperate waters of the southern hemisphere. Other genera, such as *Philinopsis*, *Chromodoris*, *Tambja*, and *Godiva*, are known to inhabit the tropical waters of the Indo-Pacific. It appears that the endemic Cape fauna originated from several sources, which indicates relationships with the Indo-Pacific tropics and the temperate faunas of both the northern and southern hemispheres. The fact that 20% of the species found on the Atlantic coast of the Cape Peninsula are also known from European waters further attests to previous influences from the northern hemisphere.

Certainly, biogeography is one of the most speculative aspects of the biology of organisms. Our understanding of the geographical affinities of the marine fauna of southern Africa is in its infancy. Each additional distributional or ecological record brings us one step closer to understanding the origins and evolution of the diversity of life which inhabits the world's oceans.

Table 2.

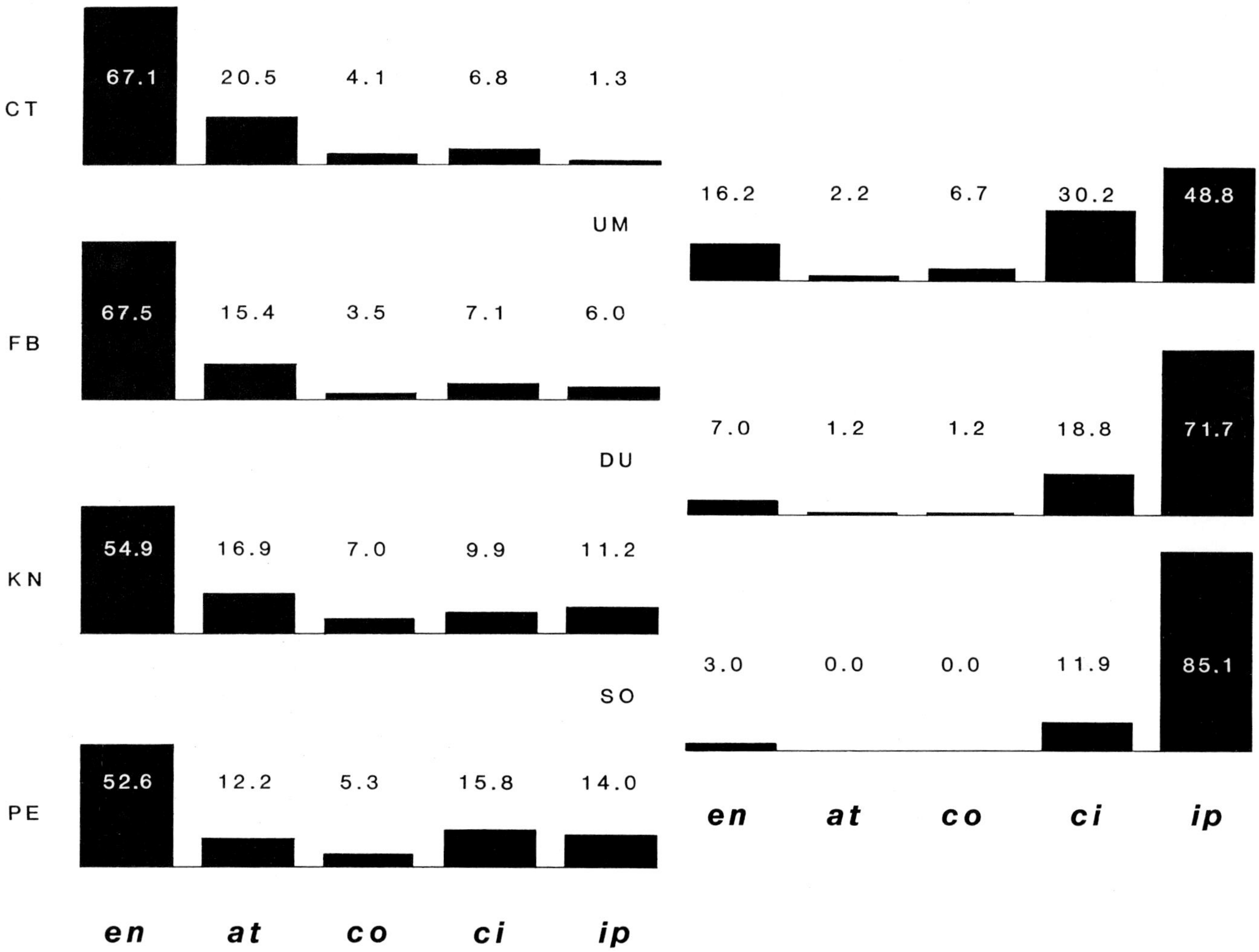

Biogeographical affinities of southern African opisthobranchs. **CT:** Cape Town; **FB:** False Bay; **KN:** Knysna; **PE:** Port Elizabeth; **UM:** Umgazana; **DU:** Durban; **SO:** Sodwana Bay; **en:** endemic; **at:** Atlantic; **co:** cosmopolitan; **ci:** circumtropical; **ip:** Indo-Pacific.

The Species of Southern African Opisthobranchia

Previously, 205 species of opisthobranchs have been recorded from the waters of southern Africa. Many of these are poorly described and were based only on preserved material. The true identity of many of these species may never be known. Of the species which have been recorded previously, living representatives of 92 species are depicted in this book. An additional 176 species are also included and represent new records of species from the region. Of these new records, 100 species are at present undescribed. Although work is underway to describe these species, they have no species name as yet. I decided that they should be included in this work to provide a more comprehensive record of the fauna present in southern Africa. Also, the inclusion of these species will hopefully stimulate divers and enthusiasts to collect additional material of these new species, many of which are known from only a single or very few specimens. In some cases, further material is absolutely necessary before these species can be named. Any specimens of these rare, undescribed species or any additional species would be greatly appreciated. Specimens and additional information can be sent to: T.M. Gosliner, Department of Invertebrate Zoology and Geology, California Academy of Sciences, Golden Gate Park, San Francisco, California 94118, U.S.A.

Table 3 lists opisthobranch species presently known from southern Africa. Species indicated in bold face are included in this work, while species designated with an asterisk (*) represent new records for southern Africa, and are reported here for the first time.

Table 3. Southern African Opisthobranch Species

Class GASTROPODA	**First record from southern Africa.**
Subclass **Opisthobranchia**	
Order Cephalaspidea	
Family Ringiculidae	
Ringicula turtoni Bartsch, 1915	Bartsch 1915
Family Acteonidae	
Acteon fortis Thiele, 1925	Thiele 1925
A. flammeus (Gmelin, 1791)	Sowerby 1897
A. pudicus (A. Adams, 1854)	Turton 1932
Rictaxis albus (Sowerby, 1873)	Sowerby 1873
Japonacteon sp.	*
Pupa affinis (A. Adams, 1854)	Sowerby 1897
P. niecaensis (Barnard, 1963)	Barnard 1963*b*
P. solidula (Linnaeus, 1758)	Sowerby 1892
P. sulcata (Gmelin, 1791)	Barnard 1963*b*
P. suturalis (A. Adams, 1854)	Sowerby 1892
P. tessellata (Reeve, 1842)	Smith 1903
Family Bullinidae	
Bullina scabra Gmelin, 1791	Bartsch 1915
B. oblonga Sowerby, 1897	Sowerby 1897
Family Hydatinidae	
Micromelo undata (Bruguière, 1792)	Sowerby 1892
Hydatina physis (Linnaeus, 1758)	Krauss 1848
H. amplustre (Linnaeus, 1758)	Sowerby 1897
H. velum (Gmelin, 1791)	Smith 1903
H. albocincta (van der Hoeven, 1811)	Macnae 1962*b*
Family Scaphandridae	
Cylichnella agulhasensis (Thiele, 1925)	Thiele 1925
C. meridionalis (Smith, 1902)	Smith 1902
C. minuscula (Turton, 1932)	Turton 1932
C. natalensis (Barnard, 1963)	Barnard 1963*b*
C. nitens (Smith, 1872)	Smith 1872
C. smithi (Bartsch, 1915)	Bartsch 1915
Cylinchna africana Bartsch, 1915	Bartsch 1915
C. bistriata Tomlin, 1920	Tomlin 1920
C. dulcis Thiele, 1925	Thiele 1925
C. nitens Smith, 1903	Smith 1903
C. remissa Smith, 1890	Turton 1932
C. tubulosa Gould, 1859	Gould 1859
Scaphander punctostriatus (Mighels, 1841)	Watson 1886
Family Retusidae	
Retusa agulhasensis Thiele, 1925	Thiele 1925
R. natalensis Barnard, 1963	Barnard 1963*b*
R. sp. cf. *nicobarica* Thiele, 1925	Barnard 1963*b*
R. sp. cf. *semen* Thiele, 1925	Barnard 1963*b*
R. truncatula (Bruguière, 1792)	Sowerby 1892
Volvulella mutabilis (Barnard, 1963)	Barnard 1963*b*
V. pia (Thiele, 1925)	Thiele 1925
V. rostrata (A. Adams, 1854)	Sowerby 1892
Family Philinidae	
Philine aperta (Linnaeus, 1767)	Linnaeus 1767
P. berghi Smith, 1910	Smith 1910
Family Gastropteridae	
Gastropteron flavobrunneum Gosliner, 1984	Gosliner 1984
G. alboaurantium Gosliner, 1984	Gosliner 1984
G. sp.	Barnard 1963*a*
Family Aglajidae	
Philinopsis cyanea (Martens, 1879)	Macnae 1962*b*
P. capensis (Bergh, 1907)	Bergh 1907
P. dubia (O'Donoghue, 1929)	O'Donoghue 1929
Chelidonura fulvipunctata Baba, 1938	*
C. hirundinina (Quoy and Gaimard, 1824)	Macnae 1962*b*
C. sp.	*
Family Haminoeidae	
Atys cylindrica (Helbling, 1779)	Martens 1879
Haminoea alfredensis Bartsch, 1915	Bartsch 1915
H. natalensis (Krauss, 1848)	Krauss 1848
Roxania utriculus (Brocchi, 1814)	Thiele 1925
Smaragdinella sieboldi A. Adams, 1864	Smith 1910
S. calyculata (Broderip and Sowerby, 1829)	*
Phaneropthalmus smaragdinus (Rüppell and Leuckart, 1831)	*
Family Runcinidae	
Metaruncina sp.	*
Family Bullidae	
Bulla ampulla (Linnaeus, 1758)	Sowerby 1892

Order Thecosomata	
Family Limacinidae	
Limacina bulimoides (Orbigny, 1836)	Pelseneer 1888
L. helicina (Phipps, 1774)	Pelseneer 1888
L. inflata (Orbigny, 1836)	Pelseneer 1888
L. lesueurii (Orbigny, 1836)	Pelseneer 1888
L. trochiformis (Orbigny, 1836)	Pelseneer 1888
Family Cavoliniidae	
Creseis acicula (Rang, 1828)	Meisenheimer 1905
C. virgula (Rang, 1828)	Meisenheimer 1905
Styliola subula (Quoy and Gaimard, 1827)	Pelseneer 1888
Clio andreae (Boas, 1886)	Meisenheimer 1905
C. chaptalii Gray, 1850	Meisenheimer 1905
C. cuspidata (Bosc,1802)	Pelseneer 1888
C. pyramidata Linnaeus, 1767	Pelseneer 1888
Cuvierina columnella (Rang, 1827)	Pelseneer 1888
Diacria quadridentata (Blainville, 1821)	Pelseneer 1888
D. trispinosa (Blainville, 1821)	Pelseneer 1888
Cavolinia gibbosa (Orbigny, 1836)	Pelseneer 1888
C. globosa (Gray, 1850)	Pelseneer 1888
C. inflexa (Lesueur, 1813)	Pelseneer 1888
C. tridentata (Niebuhr, 1775)	Pelseneer 1888
Family Peraclididae	
Peraclis moluccensis (Tesch, 1903)	Tesch 1903
P. reticulata (Orbigny, 1836)	Pelseneer 1888
Family Cymbuliidae	
Cymbulia sibogae Tesch, 1903	Tesch 1903
Gleba cordata Niebuhr, 1776	Pelseneer 1888
Family Desmopteridae	
Desmopterus papilio Chun, 1889	Meisenheimer 1905
Order Gymnosomata	
Family Clionidae	
Clione limacina (Phipps, 1774)	Pelseneer 1888
Family Pneumodermatidae	
Spongiobranchaea australis (Orbigny, 1836)	Pelseneer 1888
Order Anaspidea	
Family Akeridae	
Akera soluta (Gmelin, 1791)	Smith 1903
Family Aplysiidae	
Aplysia parvula Mörch, 1863	Sowerby 1892
A. oculifera Adams and Reeve, 1850	Macnae 1957
A. dactylomela Rang, 1828	Macnae 1955
A. maculata Rang, 1828	Rang 1828
A. juliana Quoy and Gaimard, 1832	Macnae 1955
Paraplysia lowii Gilchrist, 1900	Gilchrist 1900
Dolabella auricularia (Solander, 1786)	Krauss 1848
Family Notarchidae	
Bursatella leachi leachi (Blainville, 1817)	Krauss 1848
Bursatella leachi africana (Engel, 1927)	Engel 1927
Stylocheilus longicauda (Quoy and Gaimard, 1824)	Macnae 1962*b*
Dolabrifera dolabrifera (Rang, 1828)	Macnae 1962*b*
Order Sacoglossa	
Family Cylindrobullidae	
Ascobulla fischeri (Adams and Angas, 1864)	Sowerby 1897
Volvatella laguncula Sowerby, 1894	Sowerby 1894
Family Juliidae	
Berthelinia schlumbergeri Dautzenberg, 1895	*
Julia zebra Kawaguti, 1981	*
Family Oxynoidae	
Lobiger souverbiei Fischer, 1856	*
Lophopleurella capensis (Thiele, 1912)	Thiele 1912
Oxynoe viridis (Pease, 1861)	Sowerby 1892
Oxynoe sp.	*
Family Elysiidae	
Elysia viridis (Montagu, 1804)	Macnae 1954*b*
E. halimedae Macnae, 1954	Macnae 1954*b*
E. marginata (Pease, 1871)	Macnae and Kalk 1958
E. virgata (Bergh, 1888)	*
E. moebii (Bergh, 1888)	*
E. vatae Risbec, 1928	*
E. livida Baba, 1955	*
E. rufescens (Pease, 1871)	*
E. sp. 1	*
E. sp. 2	
E. sp. 3	*
E. sp. 4	*
E. sp. 5	*
E. sp. 6	*
E. sp. 7	*
Family Stiligeridae	
Stiliger ornatus Ehrenberg, 1831	Macnae and Kalk 1958
Placida dendritica (Alder and Hancock, 1843)	Macnae 1954*b*
Family Caliphyllidae	
Aplysiopsis sinusmensalis (Macnae, 1954)	Macnae 1954
Phyllobranchillus orientalis (Kelaart, 1858)	*
Mourgona sp. (not in key)	*
Order Notaspidea	
Family Umbraculidae	
Umbraculum sinicum (Gmelin, 1783)	Krauss 1848
Tylodina alfredensis Turton, 1932	Turton 1932
Family Pleurobranchidae	
Pleurobranchus disceptus O'Donoghue, 1929	O'Donoghue 1929
P. inhacae Macnae, 1962	Macnae 1962*a*
P. moebii Vayssière, 1898	Macnae 1962*a*
P. nigropunctatus (Bergh, 1907)	Bergh 1907
P. papillosa (O'Donoghue, 1929)	O'Donoghue 1929
P. peroni Cuvier, 1804	Macnae 1962*a*
P. perrieri Vayssière, 1896	Macnae 1962*a*
P. sculptata (O'Donoghue, 1929)	O'Donoghue 1929
P. xhosa Macnae, 1962	Macnae 1962*a*
Berthella tupala Marcus, 1957	
B. plumula (Montagu, 1803)	*
B. sp.	*
Berthellina citrina (Rüppell and Leuckart,1828)	Krauss 1848
Family Pleurobranchaeidae	
Euselenops luniceps (Cuvier, 1817)	Macnae 1962*a*
Pleurobranchaea algoensis Thiele, 1925	Thiele 1925

P. brockii Bergh, 1897	Macnae 1962*a*
P. melanopus Bergh, 1907	Bergh 1907
P. pleurobrancheana (Bergh, 1907)	Bergh 1907
P. tarda Verrill, 1880	Vayssière 1900
P. bubala Marcus and Gosliner, 1984	Marcus and Gosliner 1984
Pleurobranchella nicobarica Thiele, 1925	O'Donoghue 1929
Order Nudibranchia	
Suborder Doridacea	
Family Bathydorididae	
Doriodoxa benthalis Barnard, 1963	Barnard 1963*a*
Family Dorididae	
Doris verrucosa Linnaeus, 1758	Bergh 1907
D. sp. 1	*
D. sp. 2	*
Doriopsis pecten (Collingwood, 1881)	*
Atagema rugosa Pruvot-Fol, 1951	*
A. gibba Pruvot-Fol, 1951	*
Family Archidorididae	
Archidoris capensis Bergh, 1907	Bergh 1907
A. scripta Bergh, 1907	Bergh 1907
Alloiodoris inhacae O'Donoghue, 1929	O'Donoghue 1929
Family Discodorididae	
Discodoris coerulescens Bergh, 1888	Eliot 1910
D. fragilis (Alder and Hancock, 1864)	*
D. sp. 1	*
D. sp. 2	*
Geitodoris capensis Bergh, 1907	Bergh 1907
Anisodoris sp. 1	*
A. sp. 2	*
Thordisa burnupi Eliot, 1910	Eliot 1910
T. punctilifera Bergh, 1907	Bergh 1907
T. sp. 1	*
T. sp. 2	*
Family Asteronitidae	
Aphelodoris brunnea Bergh, 1907	Bergh 1907
A. sp.	*
Sclerodoris apiculata (Alder and Hancock, 1864)	*
S. coriacea Eliot, 1904	*
S. sp.	*
Artachaea sp.	*
Halgerda wasinensis Eliot, 1904	*
H. formosa Bergh, 1880	*
H. punctata Farran, 1902	*
Family Actinocyclidae	
Hallaxa sp.	*
Family Kentrodorididae	
Jorunna zania Marcus, 1976	*
J. tomentosa (Cuvier, 1804)	*
Gargamella sp. 1	*
G. sp. 2	*
Family Rostangidae	
Rostanga muscula (Abraham, 1877)	*
R. sp. 1	*
R. sp. 2	*

R. sp. 3	*
Family Aldisidae	
Aldisa trimaculata Gosliner, 1985	Gosliner, in Millen and Gosliner 1985
A. benguelae Gosliner, 1985	Gosliner, in Millen and Gosliner 1985
Family Platydorididae	
Platydoris scabra (Cuvier, 1806)	Smith 1903
P. cruenta (Quoy and Gaimard, 1932)	*
P. sp. (not in key)	*
Family Chromodorididae	
Chromodoris hamiltoni Rudman, 1977	*
C. africana Eliot, 1904	*
C. annulata Eliot, 1904	Eliot 1905
C. inopinata Bergh, 1905	*
C. alderi Collingwood, 1881	*
C. vicina Eliot, 1904	*
C. marginata (Pease, 1860)	*
C. geometrica (Risbec, 1928)	*
C. euelpis Bergh, 1907	Bergh 1907
C. albolimbata Bergh, 1907	Bergh 1907
C. porcata Bergh, 1888	Eliot 1910
C. sp. 1	*
C. sp. 2	*
C. sp. 3	*
C. sp. 4	*
C. sp. 5	*
C. sp. 6	*
C. sp. 7	*
C. sp. 8	*
Durvilledoris lemniscata (Quoy and Gaimard, 1832)	*
Noumea varians (Pease, 1871)	*
N. purpurea Baba, 1949	*
N. sp.	*
Thorunna horologica Rudman, 1984	*
Hypselodoris capensis (Barnard, 1927)	Barnard 1927
H. carnea (Bergh, 1889)	Eliot 1910
H. infucata (Rüppell and Leuckart, 1828)	Eliot 1905
H. maridadilus Rudman, 1977	Eliot 1905
H. sp. 1	*
H. sp. 2	*
H. sp. 3	*
Risbecia pulchella (Rüppell and Leuckart, 1828)	*
Glossodoris atromarginata (Cuvier, 1804)	*
G. sp. 1	*
G. sp. 2	*
G. sp. 3	*
G. sp. 4	*
Ceratosoma cornigerum Adams and Reeve, 1850	Macnae and Kalk 1958
C. sp.	*
Cadlina sp. 1	*
C. sp. 2	*
C. sp. 3	*
C. sp. 4	*

Family Hexabranchidae	
Hexabranchus sanguineus (Rüppell and Leuckart, 1828)	Eliot 1905
Family Dendrodorididae	
Dendrodoris nigra (Stimpson, 1855)	Barnard 1927
D. caesia (Bergh, 1907)	Bergh 1907
D. denisoni (Angas, 1864)	*
D. capensis (Bergh, 1907)	Bergh 1907
D. callosa (Bergh, 1907)	Bergh 1907
D. rubra (Kelaart, 1858)	Macnae and Kalk 1958
D. sp. (Bergh, 1907)	Bergh 1907
D. sp. 1	*
D. sp. 2	*
Doriopsilla miniata (Alder and Hancock, 1864)	*
D. capensis Bergh, 1907	Bergh 1907
D. sp. 1	*
D. sp. 2	*
Family Phyllidiidae	
Phyllidia varicosa Lamarck, 1801	*
P. sp. 1	*
P. sp. 2	*
Ceratophyllidia africana Eliot, 1903	*
Incertae sedis	
Doris natalensis Krauss, 1848	Krauss 1848
D. pseudida Bergh, 1907	Bergh 1907
D. perplexa Bergh, 1907	Bergh 1907
D. glabella Bergh, 1907	Bergh 1907
Family Onchidorididae	
Acanthodoris sp.	*
Diaphorodoris sp.	*
Family Goniodorididae	
Okenia amoenula (Bergh, 1907)	Bergh 1907
O. sp.	*
Goniodoris mercurialis Macnae, 1958	Macnae 1958
G. castanea Alder and Hancock, 1845	Macnae 1958
G. ovata Barnard, 1934	Barnard 1934
G. sp.	*
Trapania sp. 1	*
T. sp. 2	*
Ancula sp.	*
Corambe sp.	*
Family Polyceridae	
Crimora sp.	*
Thecacera pennigera (Montagu, 1804)	Barnard 1933
T. pacifica (Bergh, 1883)	Macnae 1958
T. sp.	*
Polycera capensis Quoy and Gaimard, 1824	Quoy and Gaimard 1824
P. quadrilineata (Müller, 1776)	*
P. hedgpethi Marcus, 1964	Gosliner 1982
P. sp.	*
Plocamophorus maculatus (Pease, 1860)	*
P. sp.	*
Kaloplocamus ramosus (Cantraine, 1835)	Bergh 1907
Aegires sp.	*
Limacia clavigera (Müller, 1776)	Stimpson 1854
Nembrotha livingstonei Allan, 1933	*
N. purpureolineata O'Donoghue, 1924	*
Tambja capensis (Bergh, 1907)	Bergh 1907
T. morosa (Bergh, 1877)	*
T. sp.	*
Roboastra gracilis (Bergh, 1877)	*
R. luteolineata (Baba, 1936)	*
Gymnodoris inornata (Bergh, 1880)	Macnae 1958
G. okinawae Baba, 1936	*
G. ceylonica (Kelaart, 1858)	Macnae 1958
G. alba (Bergh, 1877)	*
G. sp. 1	*
G. sp. 2	*
Lecithophorus capensis Macnae, 1958	Macnae 1958
L. sp.	*
Kalinga ornata Alder and Hancock, 1864	Bergh 1907
Family Vayssieridae	
Okadaia elegans Baba, 1931	*
Suborder Dendronotacea	
Family Bornellidae	
Bornella adamsi Gray, 1850	*
B. anguilla Johnson, 1983	*
Family Scyllaeidae	
Scyllaea pelagica Linnaeus, 1758	*
Notobryon wardi Odhner, 1936	Thompson 1981
Family Tethyidae	
Melibe rosea Rang, 1829	Rang 1829
M. pilosa Pease, 1860	Macnae and Kalk 1958
M. sp.	*
Family Dotoidae	
Doto coronata (Gmelin, 1791)	*
D. pinnatifida (Montagu, 1804)	Barnard 1927
D. rosea Trinchese, 1881	Barnard 1927
Family Marianinidae	
Marianina rosea Pruvot-Fol, 1930	*
Family Tritoniidae	
Tritonia nilsodhneri Marcus, 1983	*
T. aurantiacum Barnard, 1927	Barnard 1927
T. pallida Stimpson, 1854	Stimpson 1854
T. indecora Bergh, 1907	Bergh 1907
T. sp. 1	*
T. sp. 2	*
Tritoniodoxa capensis Bergh, 1907	Bergh 1907
Marionia sp. 1	*
M. sp. 2	*
Marioniopsis cyanobranchiata (Rüppell and Leuckart, 1831)	Eliot 1910
Suborder Arminacea	
Family Arminidae	
Armina berghi Thiele, 1925	Thiele 1925
A. capensis (Bergh, 1907)	Bergh 1907
A. euchroa (Bergh, 1907)	Bergh 1907
A. gilchristi (Bergh, 1907)	Bergh 1907
A. grisea O'Donoghue, 1929	O'Donoghue 1929
A. microdonta (Bergh, 1907)	Bergh 1907
A. natalensis (Bergh, 1866)	Bergh 1866

A. serrata O'Donoghue, 1929	O'Donoghue 1929
A. simoniana Thiele, 1925	Thiele 1925
Dermatobranchus sp. 1	*
D. sp. 2	*
D. sp. 3	*
D. sp. 4	*
Family Lemindidae	
Leminda millecra Griffiths, 1985	Griffiths 1985
Family Janolidae	
Janolus capensis Bergh, 1907	Bergh 1907
J. longidentatus Gosliner, 1981	Gosliner 1981
Bonisa nakaza Gosliner, 1981	Gosliner 1981
Suborder Aeolidacea	
Family Flabellinidae	
Flabellina funeka Gosliner and Griffiths, 1981	Gosliner and Griffiths 1981
F. capensis, (Thiele, 1925)	Thiele 1925
F. sp. 1	*
F. sp. 2	*
F. sp. 3	*
Coryphellina sp.	*
Family Eubranchidae	
Eubranchus sp. 1	*
E. sp. 2	*
E. sp. 3	*
Family Embletoniidae	
Embletonia gracilis Risbec, 1928	Gosliner and Griffiths 1981
Family Tergipedidae	
Cuthona speciosa (Macnae, 1954)	Macnae 1954*a*
C. ornata Baba, 1937	*
C. kanga (Edmunds, 1970)	*
C. annulata (Baba, 1949)	*
C. sp. 1	*
C. sp. 2	*
C. sp. 3	*
C. sp. 4	*
C. sp. 5	*
Tergipes tergipes Forskål, 1779	*
Catriona casha Gosliner and Griffiths, 1981	Gosliner and Griffiths 1981
C. columbiana (O'Donoghue, 1922)	Gosliner and Griffiths 1981
C. sp.	*
Phestilla melanobrachia Bergh, 1874	*
Family Fionidae	
Fiona pinnata (Eschscholtz, 1831)	*
Family Facelinidae	
Facelina olivacea Macnae, 1954	Macnae 1954*a*
F. annulata Macnae, 1954	Macnae 1954*a*
F. sp.	*
Pruvotfolia pselliotes (Labbé, 1923)	Barnard 1927
Moridilla brockii Bergh, 1888	*
Caloria indica (Bergh, 1896)	*
C. sp. 1	*
C. sp. 2	*
C. sp. 3	*
Favorinus japonicus Baba, 1949	*
F. ghanensis Edmunds, 1968	*
Godiva quadricolor (Barnard, 1927)	Barnard 1927
Phyllodesmium serratum (Baba, 1949)	Macnae 1954*a*
P. hyalinum Ehrenberg, 1831	*
Amanda armata Macnae, 1954	Macnae 1954*a*
Echinopsole fulvus Macnae, 1954	Macnae 1954*a*
Family Cratenidae	
Cratena capensis Barnard, 1927	Barnard 1927
C. simba Edmunds, 1970	*
C. sp. 1	*
C. sp. 2	*
C. sp. 3	*
C. sp. 4	*
Family Glaucidae	
Glaucus atlanticus Forster, 1777	Bergh 1907
Family Aeolidiidae	
Aeolidiella indica Bergh, 1888	Barnard 1927
A. alba Risbec, 1928	Gosliner 1985
Berghia chaka Gosliner, 1985	Gosliner 1985
Baeolidia palythoae Gosliner, 1985	Gosliner 1985

Identification of Opisthobranchs

The identification and classification of opisthobranch gastropods is often a difficult and painstaking task, requiring dissection of the animal and observation of the radula and reproductive system. For example, most dorid nudibranchs can be identified to generic level only after the complex arrangement of the reproductive system has been elucidated. Fortunately, once the internal morphology is known, most species can be distinguished by the external morphology of the living animal.

In order to identify opisthobranchs, based on the form of the living animal, it is necessary to understand some basic morphological terms. For this reason, an illustrated glossary has been provided.

Once the basic parts are known, opisthobranch species can be identified using the following key to the opisthobranchs. This identification key is designed to be used for living opisthobranchs; coloration is an important criterion for classification which disappears when specimens are preserved. In the case of smaller species, a hand lens or dissecting microscope may be necessary to observe the morphology of the animal.

In order to successfully use this key, the progression from one couplet to the next must be followed strictly. In each case, after selecting the more correct choice from the couplet, proceed to the couplet indicated on the right margin. Don't take any shortcuts as this can often lead one astray.

Glossary

annulate – referring to rhinophores with well separated rings (lamellae) (Fig. 7e).

bulloid – with a shell as in the bubble snail, *Bulla*, without an elevated spire (Fig. 4b, c).

caryophyllidia – notal papillae surrounded by a circle of spines (Fig. 10).

cerata (singular is ceras) – finger-like extensions from the notum which contain a branch of the digestive gland, found in the aeolid nudibranchs and in some arminaceans and sacoglossans (Fig. 7d).

convoluted crest – a sensory structure resembling a cock's comb, situated between the rhinophores in janolid nudibranchs.

digestive gland – the organ primarily responsible for digestion. In aeolids and some sacoglossans branches of the digestive gland extend into the cerata.

elevated – a shell with a well-developed spire (Fig. 4a).

extrabranchial – referring to appendages found near the gills (Fig. 8a, b).

extrarhinophoral – referring to appendages found next to rhinophores.

foot corners – the anteriormost portion of the foot; may be simply rounded, angular, or tentacular (Fig. 13).

gizzard plates – hard chitinous or calcified plates found in the esophagus of some cephalaspideans, pteropods, and anaspideans, which are used for crushing prey.

head-shield – a blunt flap of tissue on the head used by cephalaspideans for burrowing (Fig. 6a, b).

lamellae – the congested plates of annulate or perfoliate rhinophores (Fig. 7d, e).

mantle – a layer of tissue on the dorsal surface of most gastropods, which is primarily responsible for the secretion of the shell.

notum – the back surface of the animal.

oral hood – an expanded portion of the head used to trap prey in members of the genus *Melibe*.

oral tentacles – tentacles adjacent to the mouth (Fig. 6d).

papillae – elongate extensions from the body that lack lobes of the digestive gland; always longer than wide.

papillate – ornamented with numerous papillae (Fig. 7f).

parapodia – fleshy lobes extending from the body (Fig. 5).

perfoliate – rhinophores with numerous, congested rings (lamellae) (Fig. 7d).

periostracum – a brownish covering on the surface of some shells.

pinnate – branching of the gill filaments in a feather-like fashion.

pustules – irregular, slightly-elevated markings on the notum.

rachis – the central axis of the gill.

radula – a chitinous ribbon of teeth used for feeding by most mollusks.

rhinophores – paired sensory structures on the top surface of the head of most opisthobranchs (Fig. 6c, d).

rugose – rhinophores which are irregularly wrinkled (Fig. 7d).

sheath – a pit or sleeve into which the gills or rhinophores may be retracted (Fig. 7h).

siphon – a tubular canal at the posterior end of the shell (Fig. 4e).

sperm groove – a ciliated groove which is used to transport sperm to the head in most cephalaspideans and anaspideans (Fig. 5c).

tentacle – an elongate projection near the head or mouth that is generally a chemosensory organ.

tubercles – wart-like projections on the surface of the notum, which are wider than high.

velar appendages – projections from the region of the velum (Fig. 8b).

velum – the front surface of the head.

Figure 4. Shells of some opisthobranchs.

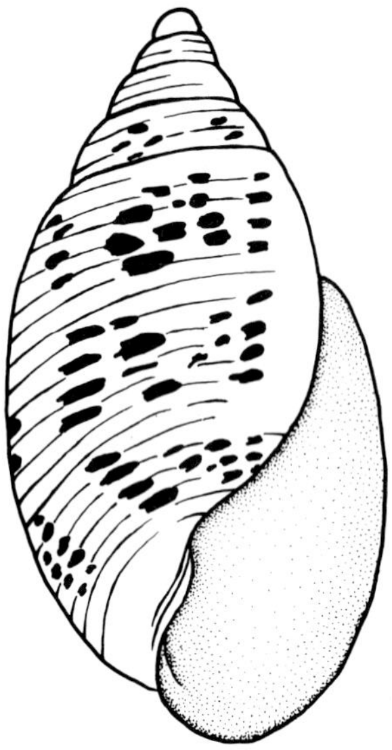

A. *Japonacteon* sp.

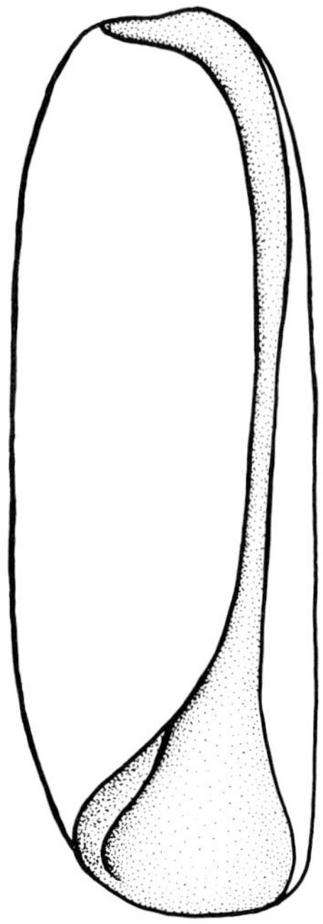

B. *Cylichna tubulosa.*

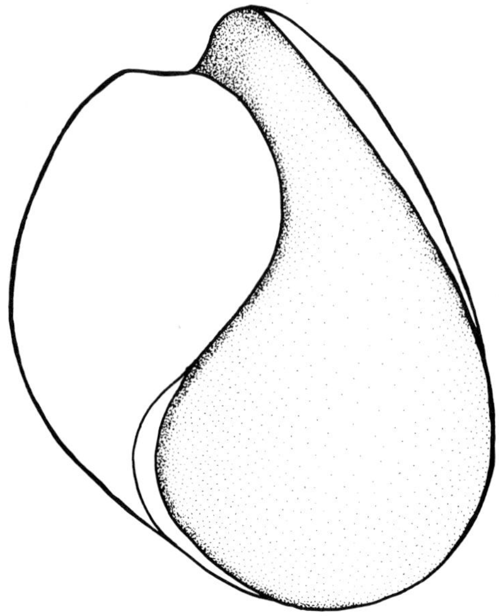

C. *Haminoea alfredensis.*

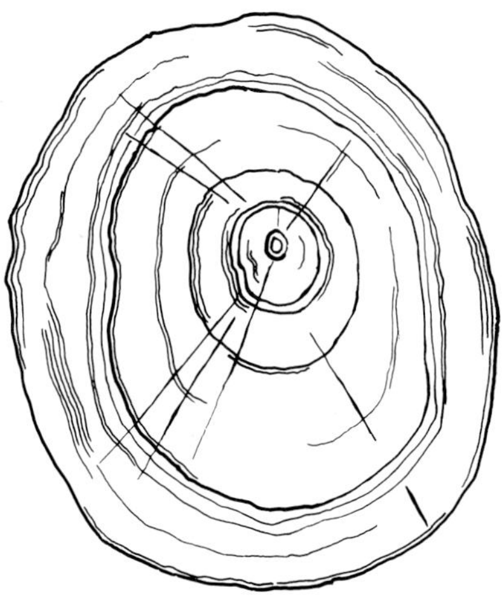

D. *Umbraculum sinicum.*

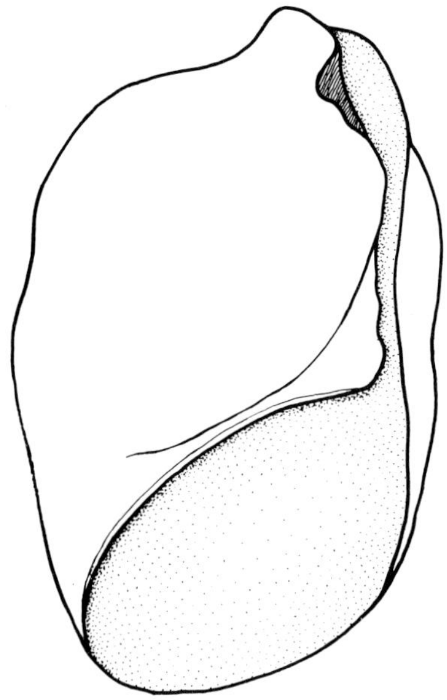

E. *Volvatella laguncula.*

Figure 5. Living animals.

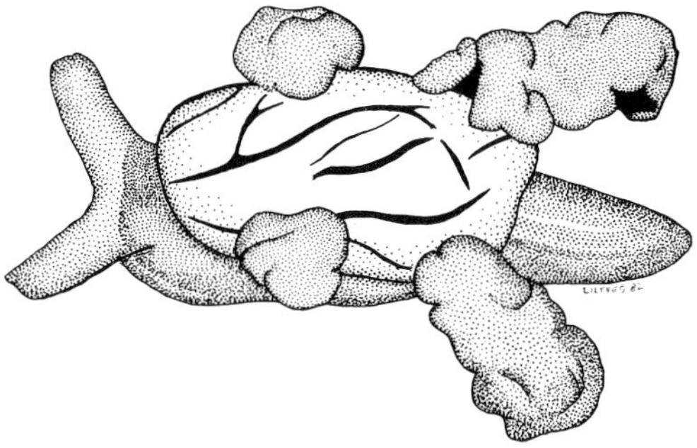

A. *Lobiger souverbiei.*

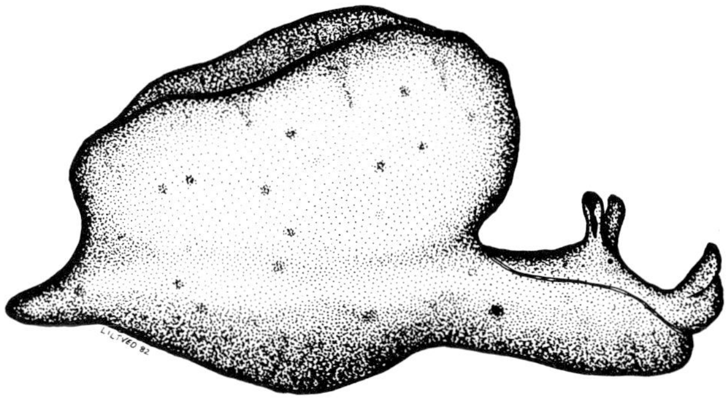

C. *Aplysia parvula.*

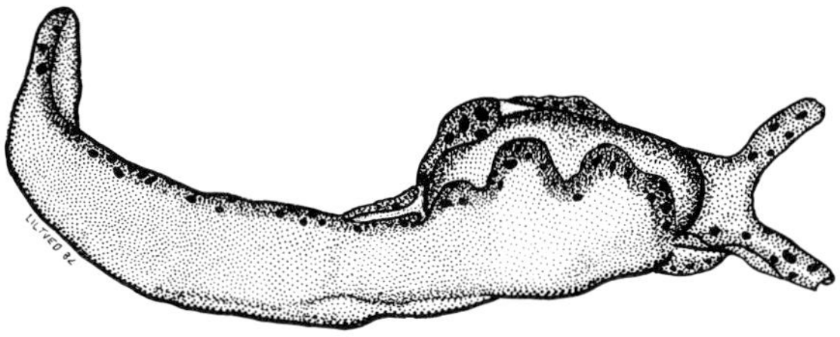

B. *Oxynoe* sp.

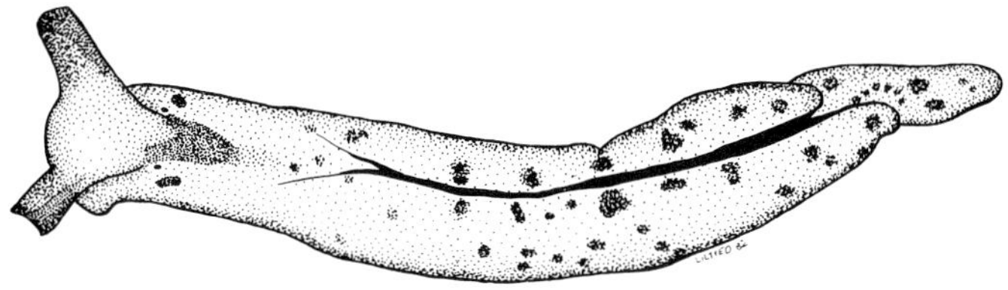

D. *Elysia viridis.*

Figure 6. Living animals.

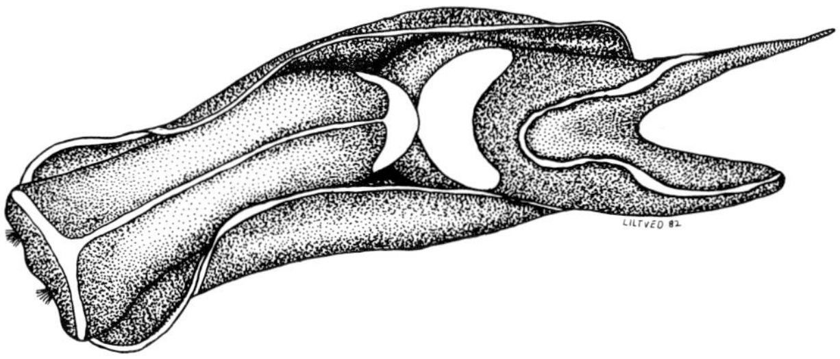

A. *Chelidonura hirundinina.*

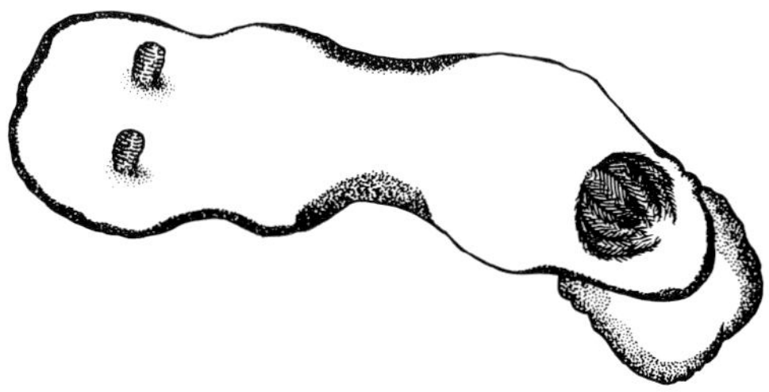

C. *Thorunna horologica.*

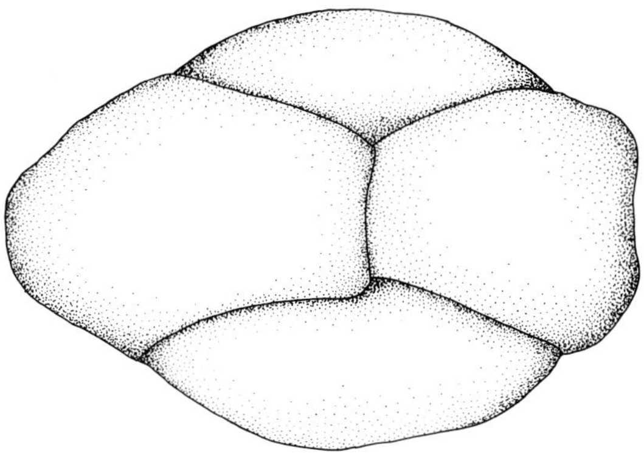

B. *Philine aperta.*

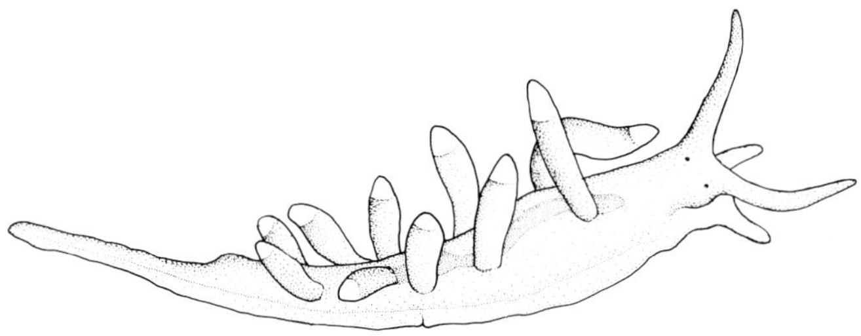

D. *Tergipes tergipes.*

Figure 7. Rhinophores.

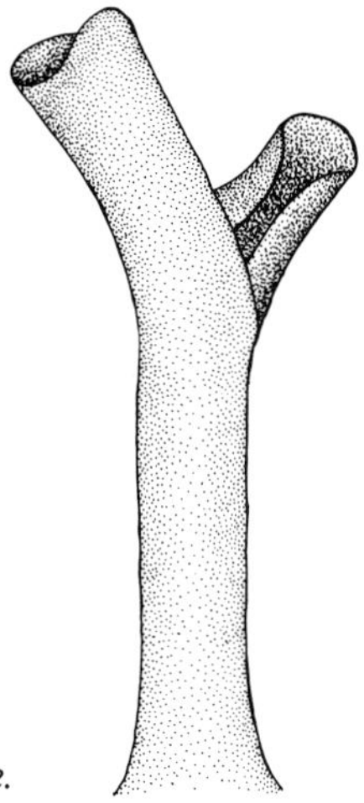

A. rolled, bifid- *Cyerce.*

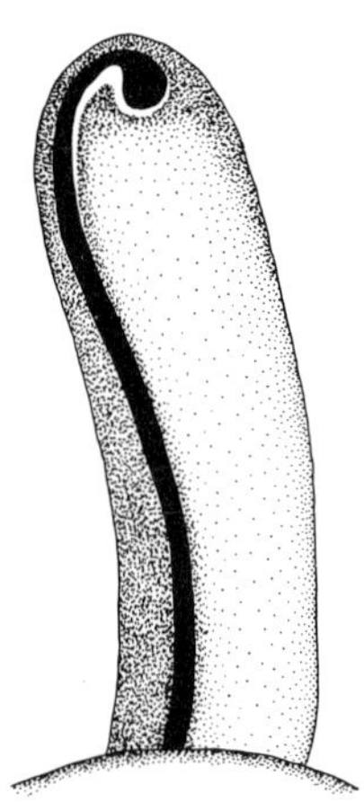

B. rolled, simple- *Berthella.*

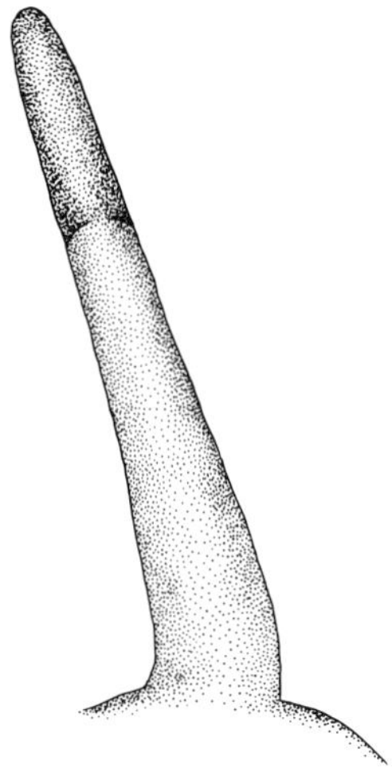

C. simple- *Cuthona.*

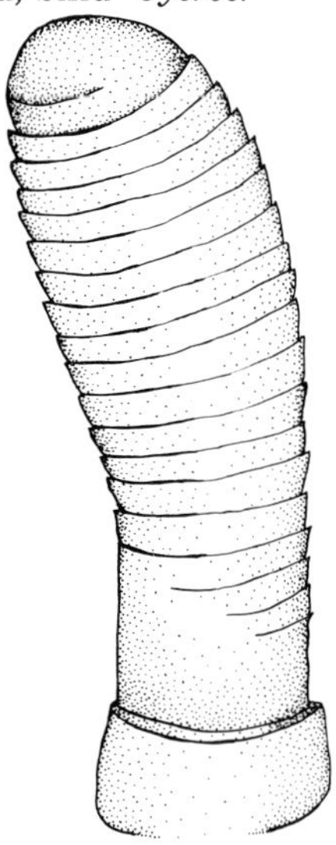

D. perfoliate- *Doris.*

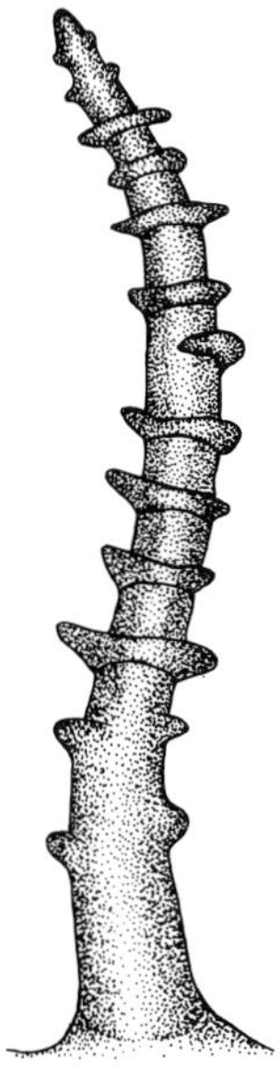

E. annulate- *Flabellina funeka.*

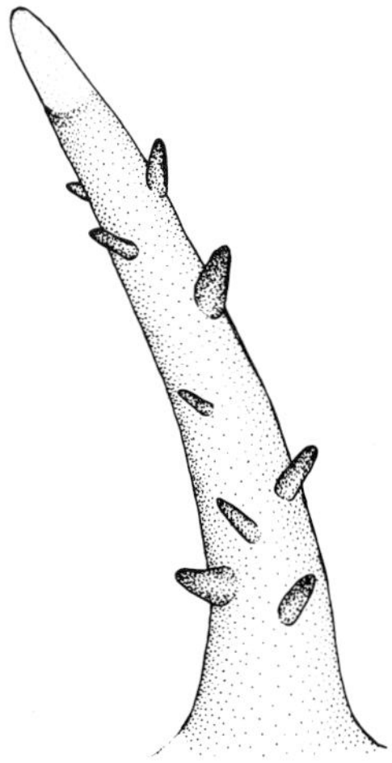

F. papillate- *Baeolidia palythoae.*

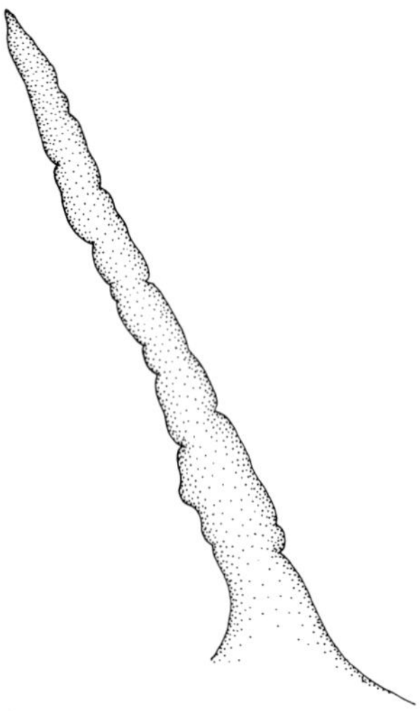

G. rugose- *Flabellina capensis.*

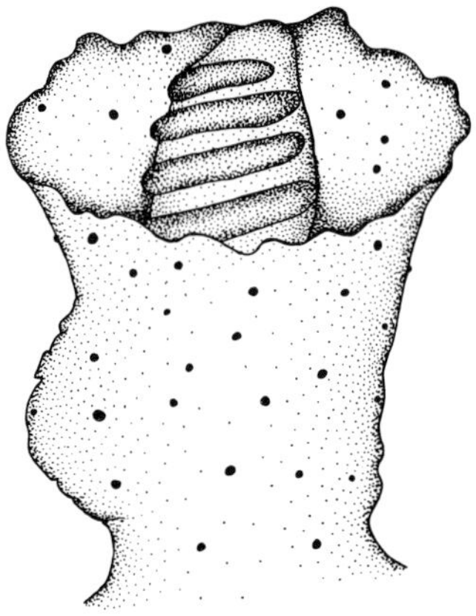

H. sheathed- *Notobryon wardi.*

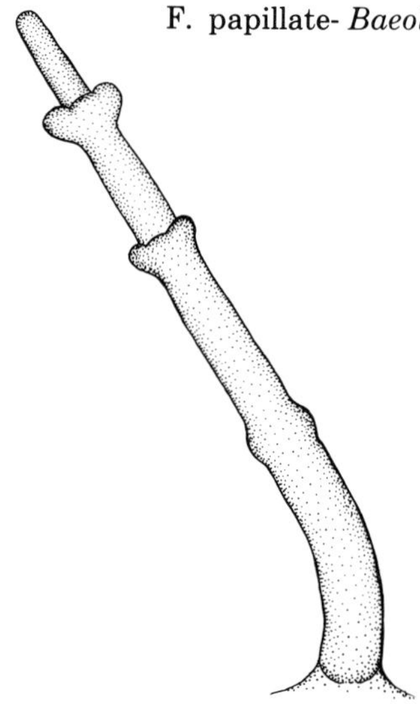

I. with swelling- *Favorinus ghanensis.*

Figure 8. Living animals.

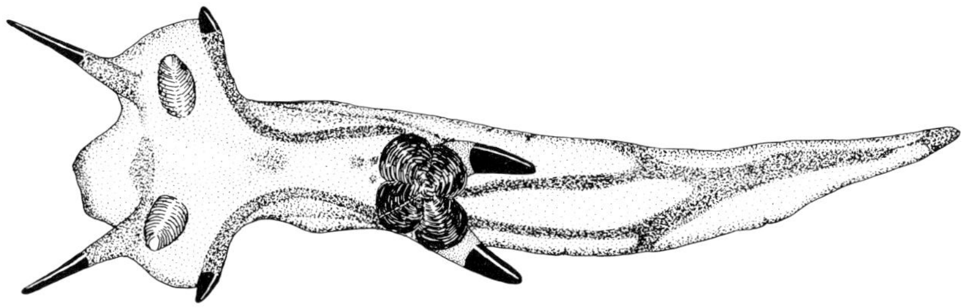

A. *Thecacera* sp.

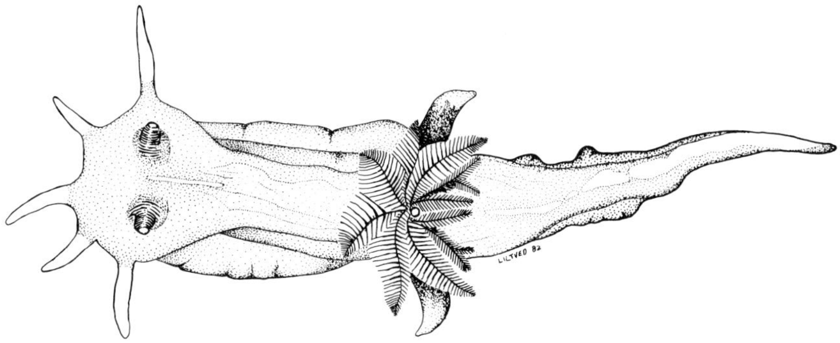

B. *Polycera capensis.*

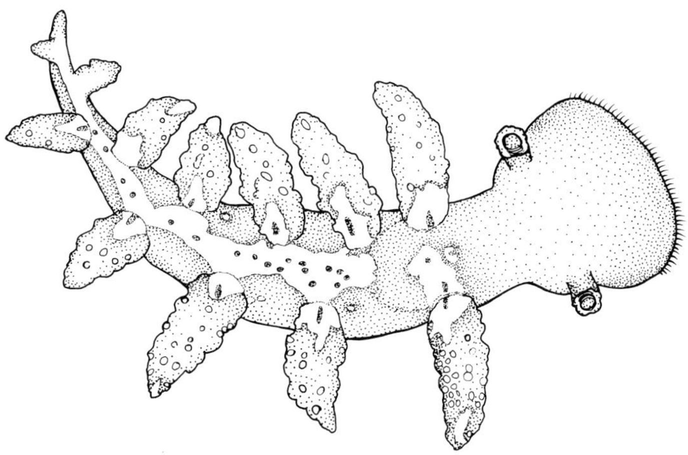

C. *Melibe rosea.*

Figure 9. Ventral view of head.

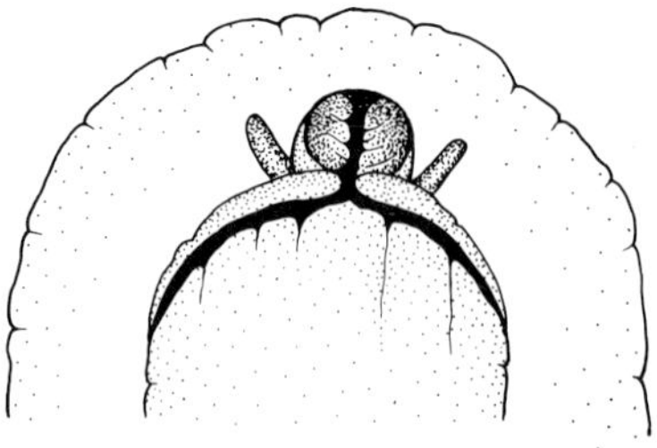

A. *Rostanga* sp. 1.

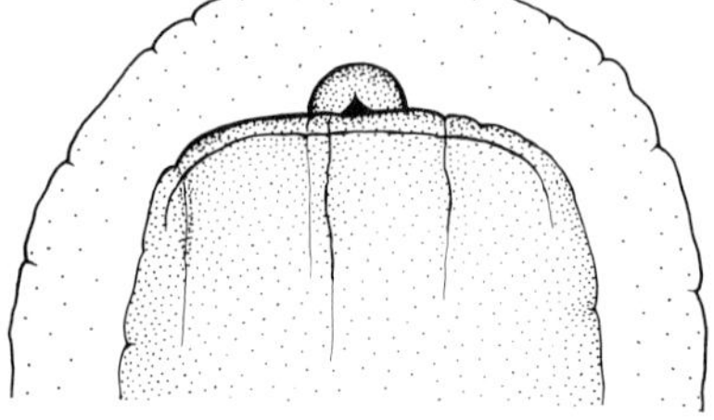

B. *Doriopsilla* sp. 2.

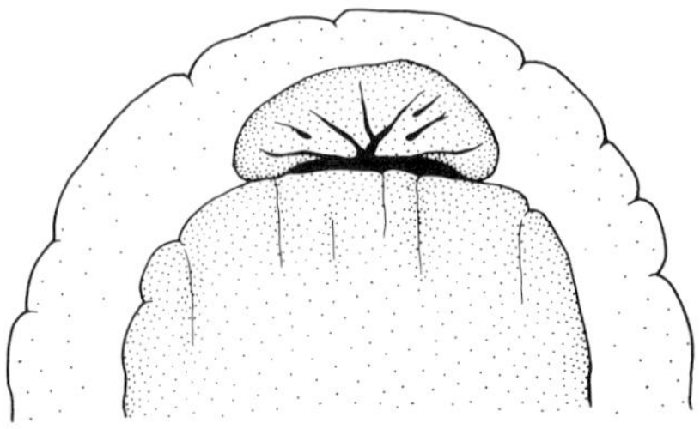

C. *Aldisa trimaculata.*

Figure 10. Tubercles.

A. *Artachaea* sp.

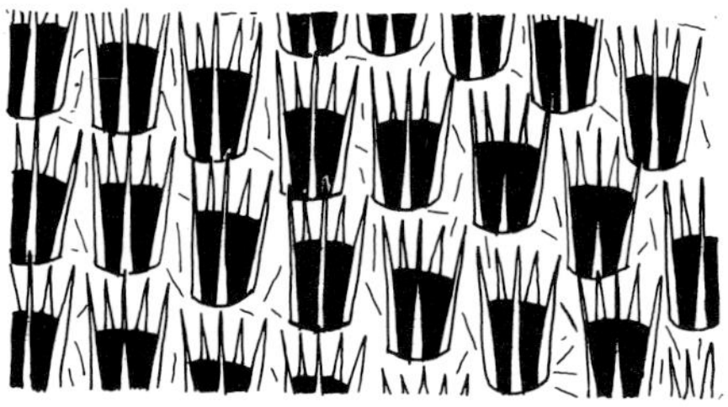

B. caryophyllidia- *Rostanga* sp. 1.

C. *Cadlina* sp. 1.

Figure 11. Gills.

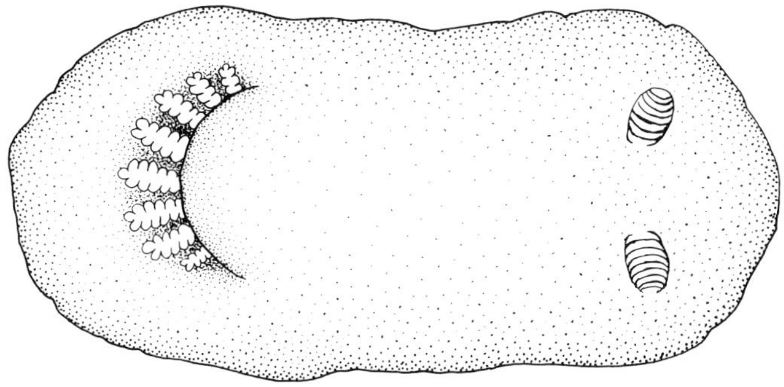

A. *Doriopsis pecten.*

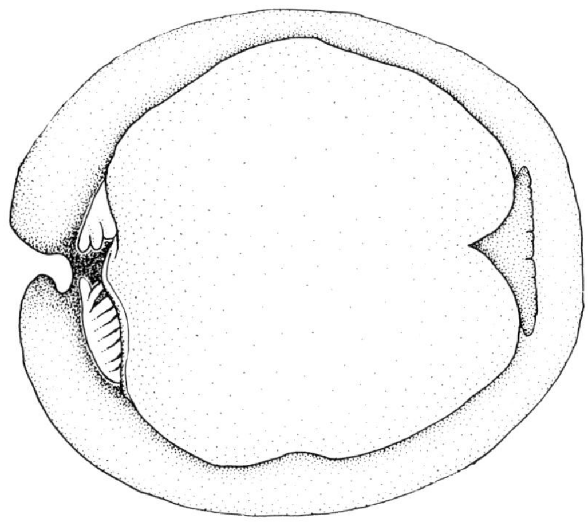

B. *Corambe* sp.

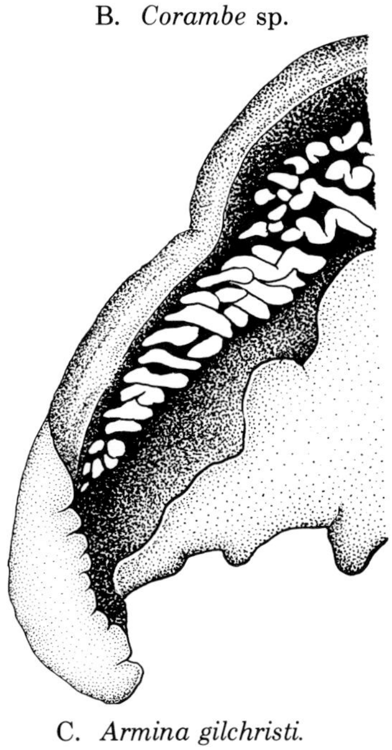

C. *Armina gilchristi.*

Figure 12. Ceratal insertion.

A. *Flabellina funeka*   B. *Cuthona speciosa.*

Figure 13. Foot corners.

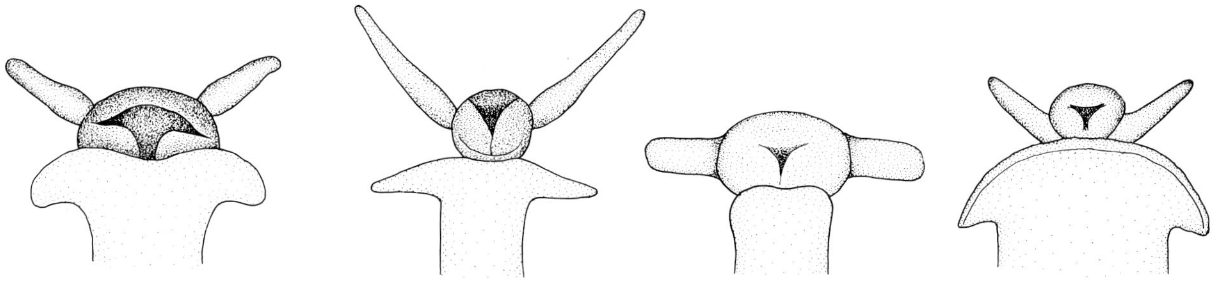

A. *Cuthona speciosa.*   B. *Flabellina funeka.*   C. *Cuthona* sp. 2.   D. *Aeloidiella indica.*

Figure 14. Cerata.

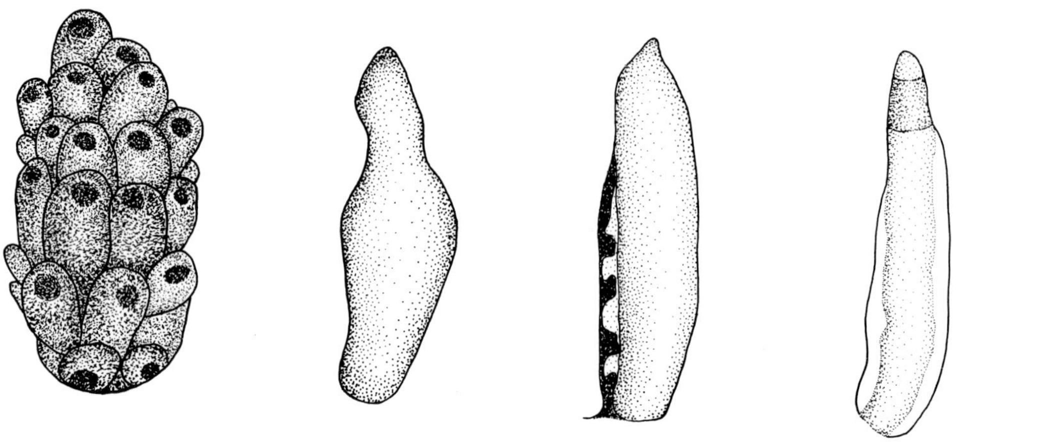

A. *Doto coronata.*   B. *Eubranchus* sp. 2.   C. *Fiona pinnata.*   D. *Flabellina funeka.*

Key To The Opisthobranchs Of Southern Africa

(Numbers in parentheses refer to species numbers in text.)

1A Shell external, may be thin and transparent ... 2
1B Shell internal or absent ... 23
1B Shell internal or absent ... 23
2A Shell bivalved ... 3
2B Shell single ... 4
3A Shell with indented posterior margin ... *Julia zebra* (34)
3B Shell without indentation ... *Berthelinia schlumbergeri* (33)
4A Shell a round, flat plate (Fig. 4d) ... *Umbraculum sinicum* (57)
4B Shell otherwise (Fig. 4a, b, c, e) ... 5
5A Shell an elongate, uncoiled cone ... *Creseis acicula* (22)
5B Shell coiled ... 6
6A Shell a small, white plate at the hind end of the animal .. *Phaneropthalmus smaragdinus* (19)
6B Shell larger, not as above ... 7
7A Spire of shell well elevated from body whorl (Fig. 4a) ... 8
7B Spire of shell depressed or bulloid (Fig. 4b, c) ... 9
8A Shell white with black squares ... *Japonacteon* sp. (2)
8B Shell white often with brown spiral lines ... *Rictaxis albus* (1)
9A Shell uniformly colored (may be transparent) or irregularly mottled ... 12
9A Shell ornamented with dark, regularly spaced lines or bands ... 10
9B Shell uniformly colored (may be transparent) or irregularly mottled ... 12
10A Shell with many fine dark brown parallel lines ... *Hydatina physis* (4)
10B Shell otherwise ... 11
11A Shell with pink or white bands separated by thin black lines ... *Hydatina amplustre* (5)
11B Shell with three black spiral lines connected by irregular black lines ... *Micromelo undata* (3)
12A Sides of shell parallel for most of their length (Fig. 4b) ... 13
12B Sides of shell not parallel (Fig. 4c) ... 14
13A Top of body whorl flat ... *Retusa truncatula* (7)
13B Top of body whorl slightly curved inwardly ... *Cylichna tubulosa* (6)
14A Shell mottled with brown and black blotches ... *Bulla ampulla* (21)
14B Shell uniformly colored (mantle may be pigmented below transparent shell) ... 15
15A Foot narrow, extending well behind shell ... 16
15B Foot wide, not extending far behind shell ... 18
16A Animal with four elongate parapodia (Fig. 5a) ... *Lobiger souverbiei* (35)
16B Animal with short, scalloped parapodia (Fig. 5b) ... 17
17A White line present on parapodial margin ... *Oxynoe* sp. (37)
17B Parapodial margin same color as body ... *Oxynoe viridis* (36)
18A Body much narrower in front, animal swims with parapodia ... *Akera soluta* (23)
18B Body narrower in rear, animal unable to swim ... 19
19A Body uniformly cream or white ... 20
19B Body brown or green with variously colored spots ... 21
20A Shell bulloid ... *Ascobulla fischeri* (31)
20B Shell with elongate siphon at posterior end (Fig. 4e) ... *Volvatella laguncula* (32)
21A Parapodia extend to hind end of animal ... *Smaragdinella calyculata* (18)
21B Parapodia extend only to middle of shell ... 22
22A Hind end of head-shield deeply bilobed ... *Haminoea natalensis* (17)
22B Hind end of head-shield only slightly indented ... *Haminoea alfredensis* (16)
23A Animal with well developed head-shield, rhinophores absent (Fig. 6a, b) ... 24
23B Animal without head-shield, rhinophores present (Fig. 6c, d) ... 32
24A Posterior end of head-shield forming a siphon ... 25
24B Posterior end of head-shield not forming a siphon ... 26
25A Animal yellow with brown spots ... *Gastropteron flavobrunneum* (14)
25B Animal pinkish white with orange siphon ... *Gastropteron alboaurantium* (15)
26A Body cream or white ... 27

26A Body cream or white 27
26B Body green, brown, or black 28
27A Hind end of body blunt, "tails" absent .... *Philine aperta* (8)
27B Hind end of body with paired, short "tails" .... *Chelidonura* sp. (13)
28A Animal small, uniformly green or black .... *Metaruncina* sp. (20)
28B Animal with mottling, spots, or other markings 29
29A Body with elongate, tapered "tails" (Fig. 6a) 30
29B Body with short, blunt "tails" (Fig. 6b) 31
30A Head with white "W"-shaped marking .... *Chelidonura fulvipunctata* (12)
30B Head with white or blue "T"-shaped marking .... *Chelidonura hirundinina* (11)
31A Head-shield with mid-dorsal line .... *Philinopsis capensis* (10)
31B Head-shield without mid-dorsal line .... *Philinopsis cyanea* (9)
32A Rhinophores rolled (Fig. 7a, b) 33
32B Rhinophores solid (Fig. 7c-i) 67
33A Parapodia present, may be short (Fig. 5) 34
33B Parapodia absent 55
34A Sperm groove present (Fig. 5c) 35
34B Sperm groove absent (Fig. 5d) 41
35A Body covered by papillae 36
35B Body smooth 38
36A Hind end of animal ending abruptly in large disc .... *Dolabella auricularia* (28)
36B Foot extending well behind body 37
37A Brown longitudinal lines present on body .... *Stylocheilus longicauda* (30)
37B Brown lines absent .... *Bursatella leachi* (29)
38A Posterior end of foot with distinct sucker .... *Aplysia juliana* (27)
38B Posterior end of foot without sucker 39
39A Clusters of opaque white spots, black circles absent .... *Aplysia parvula* (24)
39B Black circles present 40
40A Black circles more than 3 mm in diameter .... *Aplysia dactylomela* (26)
40B Black circles less than 2 mm in diameter .... *Aplysia oculifera* (25)
41A Parapodial margin different color than body 42
41B Parapodial margin same color as body 49
42A Tips of rhinophores with red or orange pigment, body bluish 43
42B Tips of rhinophores not red or orange, body not bluish 44
43A Black spots present on parapodia .... *Elysia vatae* (43)
43B Opaque white spots present on parapodia .... *Elysia* sp. 2 (47)
44A Animal blue-green, parapodial margin with burnt orange and black .... *Elysia moebii* (42)
44B Animal not as above 45
45A Body with opaque white and black spots, parapodial margins orange and black .... *Elysia marginata* (40)
45B Body with blue or purple pigment 46
46A Outer parapodial margin purple .... *Elysia rufescens* (45)
46B Outer parapodial margin orange, yellow, or black 47
47A Rhinophores and head black .... *Elysia livida* (44)
47B White pigment present on rhinophores and/or head 48
48A Longitudinal undulating blue line present near parapodial margin .... *Elysia* sp. 3 (48)
48B Isolated blue spots present on parapodia .... *Elysia* sp. 1 (46)
49A Body with black longitudinal lines .... *Elysia virgata* (41)
49B Black lines absent from body 50
50A Body with tubercles 51
50B Body without tubercles 54
51A Rhinophores tuberculate .... *Elysia* sp. 5 (50)
51B Rhinophores smooth 52
52A Yellow spots absent, blue pigment present .... *Elysia* sp. 4 (49)
52B Yellow spots present on parapodia 53
53A Yellow spots scattered throughout parapodia .... *Elysia* sp. 6 (51)

53B Yellow spots present only around parapodial margin .... *Elysia* sp. 7 (52)
54A Body stout, hind end of foot rounded .... *Elysia halimedae* (39)
54B Body slender, hind end of foot pointed .... *Elysia viridis* (38)
55A Cerata present 56
55B Cerata absent 58
56A Rhinophores bifid (Fig. 7A) .... *Phyllobranchillus orientalis* (55)
56B Rhinophores unbranched (Fig. 7B) 57
57A Digestive gland visible in notum, head, and cerata .... *Placida dendritica* (53)
57B Digestive gland not visible .... *Aplysiopsis sinusmensalis* (54)
58A Rhinophores well separated 59
58B Rhinophores joined closely together 60
59A Body with opaque white pustules .... *Pleurobranchaea bubala* (67)
59B Body without opaque white pustules .... *Pleurobranchaea tarda* (66)
60A Body color white to cream 61
60B Body red, orange, or plum 63
61A Body uniformly cream .... *Berthella plumula* (63)
61B Body with white pigment 62
62A Body uniformly white .... *Berthella* sp. (64)
62B Body with mid-dorsal opaque white markings .... *Berthella tupala* (62)
63A Body smooth .... *Berthellina citrina* (65)
63B Body with reticulations and/or tubercles 64
64A Body without tubercles .... *Pleurobranchus xhosa* (61)
64B Body with tubercles 65
65A Body with elongate tubercles .... *Pleurobranchus nigropunctatus* (59)
65B Body with low, rounded tubercles 66
66A Many small tubercles present on back .... *Pleurobranchus peroni* (60)
66B Few large tubercles present on back .... *Pleurobranchus inhacae* (58)
67A Circle of gills present on hind end of back (Figs. 6c, 8a, b) 68
67B Gills of different form or absent (Figs. 6d, 8c, 11a, b, c) 186
68A No appendages extending from velum, rhinophores, notum, or gills 69
68B Appendages present on velum, rhinophores, notum, or gills (Fig 8a, b) 168
69A Gills completely retractile into sheath 70
69B Gills may contract slightly, sheath absent 151
70A Head well developed, round or semicircular when viewed ventrally (Fig. 9a, c) 71
70B Head not well developed when viewed ventrally (Fig. 9b) 144
71A Back with tubercles, papillae or ridges, body rigid 72
71B Back smooth, or if tubercles present, body is soft and fleshy 106
72A Back with distinct ridges 73
72B Back without distinct ridges 77
73A Body with single, mid-dorsal ridge .... *Atagema rugosa* (72)
73B Body with interconnecting ridges 74
74A Body black with yellow-orange ridges .... *Halgerda wasinensis* (88)
74B Body white or orange 75
75A Body orange with white ridges .... *Sclerodoris* sp. (86)
75B Body white with yellow or orange ridges 76
76A Black or plum pigment forming spots and lines throughout dorsum .... *Halgerda formosa* (89)
76B Black or plum pigment forming spots only nearly margin .... *Halgerda punctata* (90)
77A Notum with caryophyllidia (Fig. 10b) 78
77B Notum with tubercles or papillae but without caryophyllidia (Fig. 10a, c) 87
78A Body yellowish, orange, or red 79
78B Body white, grey, or brown 82
79A Body with black spots .... *Rostanga* sp. 1 (97)
79B Body without black spots 80
80A Body light orange with darker orange circles surrounded by white .... *Anisodoris* sp. 2 (79)
80B Body without darker orange circles 81
81A Body with white blotches .... *Rostanga* sp. 3 (99)

81B Body with brown blotches. . . . . *Rostanga muscula* (96)
82A Body grey with darker spots 83
82B Body white or brown, usually with brown or black spots. 84
83A Body with darker grey spots. . . . . *Jorunna tomentosa* (93)
83B Body with black spots surrounded by white rings . . . . *Gargamella* sp. 1(94)
84A Body brown with large papillae . . . . *Sclerodoris coriacea* (85)
84B Body white. 85
85A Body with brown spots . . . . *Gargamella* sp. 2 (95)
85B Body with black spots. 86
86A Body with black spots, gills white . . . . *Rostanga* sp. 2(98)
86B Body with clusters of black tubercles, gills with black pigment. . . . . *Jorunna zania* (92)
87A Body with elongate fleshy papillae extending from tubercles 88
87B Body with regular wart-like tubercles 90
88A Body greenish. . . . . *Thordisa* sp. 1 (80)
88B Body yellow or yellow-brown. 89
89A Body lemon yellow with brownish papillae . . . . *Thordisa* sp. 2 (81)
89B Body yellow-brown with black spots . . . . *Sclerodoris apiculata* (84)
90A Body white, tubercles small. 91
90B Body not white, variously colored 93
91A Body without marginal glands that appear as isolated tubercles . . . . *Discodoris* sp. 1 (75)
91B Body with yellow or white glands near outer margin of notum. 92
92A Body with brown mid-dorsal mottling . . . . *Cadlina* sp. 1 (139)
92B Body without brown mottling. . . . . *Cadlina* sp. 2 (140)
93A Head semicircular, without tentacles ventrally (Fig. 9c). 94
93B Head circular, with ventral tentacles (Fig. 9a) 95
94A Body red-orange . . . . *Aldisa benguelae* (101)
94B Body cream to tan with three (rarely two) brown spots . . . . *Aldisa trimaculata* (100)
95A Rhinophore sheaths elevated, conical . . . . *Atagema gibba* (73)
95B Rhinophore sheaths at same level as notum 96
96A Tentacles triangular with a groove. 97
96B Tentacles rounded, finger-like 102
97A Body yellow or yellow-brown. 98
97B Body not yellow or yellowish. 101
98A Body with small tubercles with opaque white tips . . . . *Hallaxa* sp. (91)
98B Body with large wart-like tubercles. 99
99A All tubercles yellow. 100
99B Most tubercles dark brown or black . . . . *Doris* sp. 2 (70)
100A Body uniformly yellow or yellow brown . . . . *Doris verrucosa* (68)
100B Body with dark brown spots . . . . *Doris* sp. 1 (69)
101A Body brown with darker brown "W" shaped marking. . . . . *Cadlina* sp. 4 (142)
101B Body bluish with darker blue or black spots . . . . *Cadlina* sp. 3 (141)
102A Body pinkish-grey with rows of brown spots. . . . . *Discodoris fragilis* (74)
102B Body brown or orange 103
103A Black ring present around base of tubercles, foot with black spots. . . . . *Artachaea* sp. (87)
103B Black rings absent, foot without black spots. 104
104A Body orange with brown spots. . . . . *Geitodoris capensis* (77)
104B Body brown. 105
105A Some tubercles purple. . . . . *Discordoris* sp. 2 (76)
105B All tubercles brown, with darker brown spots on notum . . . . *Anisodoris* sp. 1 (78)
106A Edge of notum permanently convoluted 107
106B Edge of notum straight or temporarily convoluted 111
107A Outer notal margin grey-blue, notum mottled with brown and white spots. . . . .
. . . . *Glossodoris* sp. 3 (136)
107B Outer notal margin not grey-blue. 108
108A Notum whitish 109

108A Notum whitish 109
108B Notum brown, green, or reddish 110
109A Notal margin white .... *Glossodoris* sp. 4 (137)
109B Notal margin yellow .... *Glossodoris* sp. 1 (134)
110A Outer margin of notum black with inner yellow margin .... *Glossodoris* sp. 2 (135)
110B Outer margin purple, inner margin absent .... *Glossodoris atromarginata* (133)
111A Body with longitudinal lines or stripes on notum 112
111B Body without longitudinal lines or stripes 121
112A One or more longitudinal white lines or stripes present 113
112B Longitudinal lines or stripes not white 120
113A Red or yellow longitudinal lines present in addition to white 114
113B Only white lines present 115
114A Lateral yellow stripe present on either side of central white stripe .... *Durvilledoris lemniscata* (120
114B Red lines alternating with white lines .... *Hypselodoris* sp. 1 (129)
115A Body black .... *Chromodoris africana* (105)
115B Body white, pink, or purple 116
116A Outer marginal line of notum white or cream 117
116B Outer marginal line of notum not white or cream 118
117A Marginal line cream, undulating .... *Chromodoris* sp. 5 (116)
117B Marginal line opaque white, not undulating .... *Noumea purpurea* (122)
118A Marginal line orange, continuous .... *Chromodoris* sp. 4(115)
118B Marginal line purple, discontinuous 119
119A Body pinkish brown, marginal line discontinuous anteriorly .... *Hypselodoris capensis* (125)
119B Body white, marginal line continuous anteriorly .... *Hypselodoris carnea* (126)
120A Longitudinal lines purple .... *Hypselodoris maridadilus* (128)
120B Longitudinal lines black .... *Chromodoris hamiltoni* (104)
121A Body flat and rigid .... *Platydoris cruenta* (102)
121B Body not flat or rigid 122
122A Body orange or red, margins of notum folded, may unfold when disturbed .... *Hexabranchus sanguineus* (143)
122B Margins of notum never folded 123
123A Body white with a few large dark brown spots .... *Aphelodoris* sp. (83)
123B Color not as above 124
124A Body white with dark brown to black mottling .... *Aphelodoris brunnea* (82)
124B Color not as above 125
125A Yellow spots present on notum 126
125B Yellow spots absent 132
126A Gills and rhinophores red 127
126B Gills and rhinophores with blue or purple pigment 130
127A Body with network of dark purple lines .... *Chromodoris geometrica* (111)
127B Body with black spots 128
128A Notum largely dark blue and yellow with large black spots .... *Hypselodoris infucata* (127)
128B Notum largely whitish with minute black spots 129
129A Middle of notum pinkish .... *Hypselodoris* sp. 2 (130)
129B Middle of notum opaque white .... *Hypselodoris* sp. 3 (131)
130A Black spots present on notum .... *Chromodoris vicina* (109)
130B Black spots absent 131
131A Two large blue circles present around gills and rhinophores .... *Chromodoris annulata* (106)
131B Blue circles absent, yellow spots present on foot .... *Risbecia pulchella* (132)
132A Body with reticulate pattern of red on all or part of notum 133
132B Reticulate pattern of red absent 135
133A Reticulate pattern restricted to central portion of notum .... *Chromodoris alderi* (108)
133B Reticulate pattern covering most of notum 134
134A Margin of notum with outer orange band and inner band of purple with opaque white spots .... *Chromodoris inopinata* (107)

134B Margin of notum with orange band only .... *Chromodoris* sp. 6 (117)
135A Notum white with large blue spots .... *Ceratosoma* sp. (138)
135B Notum without large blue spots 136
136A Black or dark purple spots present on notum, may be restricted to margin 137
136B Black or purple spots absent 139
137A Notum and rhinophores yellow, black spots large .... *Chromodoris* sp. 3 (114)
137B Notum white 138
138A Rhinophores brown .... *Chromodoris* sp. 8 (119)
138B Rhinophores and gills orange with opaque white spots .... *Chromodoris* sp. 7 (118)
139A Notum pink or purple 140
139B Notum opaque white 141
140A Margin of notum yellow, undulating .... *Noumea varians* (121)
140B Notum pink with opaque white glands on margins .... *Chromodoris* sp. 2 (113)
141A Rhinophores white .... *Chromodoris* sp. 1 (112)
141B Rhinophores not white 142
142A Rhinophores brown with opaque white lines, gills white .... *Chromodoris marginata* (110)
142B Rhinophores and gills red to orange 143
143A Foot uniformly white .... *Noumea* sp. (123)
143B Foot with blue marginal line .... *Thorunna horologica* (124)
144A Notum smooth 145
144B Notum with tubercles (may be small) 147
145A Body orange to black .... *Dendrodoris nigra* (144)
145B Body pinkish, white or green 146
146A Notum greenish .... *Dendrodoris* sp. 1 (147)
146B Notum white to pink with numerous blue spots .... *Dendrodoris caesia* (145)
147A Body soft with few tubercles 148
147B Body granular, usually with many small tubercles 149
148A Rhinophores and gills red, tubercles pink .... *Dendrodoris* sp. 2 (148)
148B Rhinophores and gills brown, blue spots present .... *Dendrodoris denisoni* (146)
149A Body uniformly bluish white to brown .... *Doriopsilla* sp. 2 (151)
149B Body not uniformly colored 150
150A Body white with black spots .... *Doriopsilla* sp. 1 (150)
150B Body orange or yellow with opaque white spots or lines .... *Doriopsilla miniata* (149)
151A Body mostly black 152
151B Body variously colored, black lines may be present 156
152A Body without longitudinal lines, green or blue notal margin present 153
152B Body with longitudinal lines 154
153A Continuous green line forming margin of notum .... *Tambja capensis* (182)
153B Interrupted blue-green line forming margin .... *Tambja morosa* (183)
154A Longitudinal lines green .... *Tambja* sp. (184)
154B Longitudinal lines yellow or gold 155
155A Longitudinal lines extend onto gills .... *Roboastra luteolineata* (186)
155B Longitudinal lines terminate before gills .... *Roboastra gracilis* (185)
156A Body without thickened border of notum 157
156B Body with border of notum thickened 166
157A Gills and rhinophores bright crimson .... *Nembrotha purpureolineata* (181)
157B Gills and rhinophores variously colored, not crimson 158
158A Body reddish with large orange spots .... *Nembrotha livingstonei* (180)
158B Body white, yellow, or orange 159
159A Distinct notal ridge present along most of body 160
159B Notal ridge absent or present only around head 162
160A Body white 161
160B Body orange .... *Lecithophorus* sp. (193)
161A Body with papillae and orange marginal line .... *Diaphorodoris* sp. (157)
161B Body without papillae or orange line .... *Lecithophorus capensis* (192)

162A Body yellow to orange .... *Gymnodoris inornata* (187)
162B Body white to cream 163
163A Body with tubercles .... *Gymnodoris okinawae* (188)
163B Body smooth 164
164A Body short, stout .... *Gymnodoris* sp. 2 (191)
164B Body slender, elongate 165
165A Gills forming only a semicircle around anus .... *Gymnodoris* sp. 1 (190)
165B Gills almost forming a complete circle around anus .... *Gymnodoris alba* (189)
166A Body dark brown with yellow spots,mid-dorsal ridge present .... *Goniodoris castanea* (161)
166B Body white to tan, mid-dorsal ridge absent 167
167A Body white to tan with opaque white spots .... *Goniodoris mercurialis* (160)
167B Body opaque white with brown rhinophores .... *Goniodoris* sp. (162)
168A Elongate papillae present on notum 169
168B Elongate papillae absent, although extra-rhinophoral or extra-branchial process may be present 176
169A Papillae uniformly scattered over notum 170
169B Papillae restricted to notal edge or mid-dorsal line 171
170A Body brown, papillae simple .... *Acanthodoris* sp. (156)
170B Body orange, papillae bluish, branched .... *Crimora* sp. (167)
171A Papillae branched 172
171B Papillae simple 174
172A Papillae longer than gills .... *Kaloplocamus ramosus* (177)
172B Papillae shorter than gills 173
173A Body cream with yellow spots .... *Plocamophorus maculatus* (175)
173B Body brown with opaque white spots .... *Plocamophorus* sp. (176)
174A Papillae elongate, absent from mid-dorsal line .... *Okenia amoenula* (158)
174B Papillae present along mid-dorsal line 175
175A Single mid-dorsal papilla present in front of gills .... *Okenia* sp. (159)
175B Numerous mid-dorsal papillae present .... *Limacia clavigera* (179)
176A Extra-rhinophoral collar or papillae present (Fig. 8a) 177
176B Extra-rhinophoral collar or papillae absent (Fig. 8b) 183
177A Notum with numerous papillae .... *Aegires* sp. (178)
177B Notum without papillae other than extrabranchial papillae 178
178A Extrarhinophoral process a simple collar without papillae 179
178B Extrarhinophoral processes include elongate papillae (Fig. 8a) 180
179A Body uniformly orange with blue and black markings .... *Thecacera pacifica* (169)
179B Body white with orange and black spots .... *Thecacera pennigera* (168)
180A Body with opaque white lines .... *Trapania* sp. 1 (163)
180B Body without opaque white lines 181
181A Body brown, extra-rhinophoral papillae yellow-orange .... *Trapania* sp. 2(164)
181B Body whitish or grey 182
182A Body with yellow-orange lines .... *Thecacera* sp. (170)
182B Body with brown patches .... *Ancula* sp. (165)
183A Extrabranchial process single per side (Fig. 8b) 184
183B Extrabranchial processes two or three per side 185
184A Yellow lines present on body .... *Polycera capensis* (171)
184B Yellow spots present on body .... *Polycera quadrilineata* (172)
185A Body greenish without longitudinal lines .... *Polycera hedgpethi* (173)
185B Body white with black lines and yellow spots .... *Polycera* sp. (174)
186A Gills and cerata absent from notum 187
186B Gills or cerata present along margin of notum 199
187A Gills present in some form 188
187B Gills entirely absent 194
188A Gills pectinate, mid-dorsal on notum (Fig. 11a) .... *Doriopsis pecten* (71)
188B Gills not present on notum 189
189A Gills situated within posterior notch (Fig. 11b) .... *Corambe* sp. (166)

189A Gills situated within posterior notch (Fig. 11b) .... *Corambe* sp. (166)
189B Gills situated along sides between mantle and foot (Fig. 11c) 190
190A Anus present on right side of body .... *Armina gilchristi* (210)
190B Anus along mid-dorsal line 191
191A Notal tubercles scattered, mushroom-shaped .... *Ceratophyllidia africana* (155)
191B Notal tubercles in distinct rows, simply rounded 192
192A Body blue with yellow tubercles and black lines .... *Phyllidia varicosa* (152)
192B Body pink and black 193
193A Central transverse line disrupting longitudinal rows of tubercles .... *Phyllidia* sp. 1 (153)
193B Longitudinal rows of tubercles continuous except central row .... *Phyllidia* sp. 2 (154)
194A Animal minute, less than 5 mm in length, uniformly red-orange .... *Okadaia elegans* (194)
194B Animal longer than 5 mm in length, variously colored 195
195A Margin of body highly convoluted .... *Leminda millecra* (215)
195B Body with numerous longitudinal ridges 196
196A Notal ridges interrupted .... *Dermatobranchus* sp. 2 (212)
196B Notal ridges continuous throughout length of animal 197
197A Body brown, pink to purple, ridges white 198
197B Body and ridges white, brown spots present .... *Dermatobranchus* sp. 3 (213)
198A Notum without brown mottling .... *Dermatobranchus* sp. 1 (211)
198B Notum with brown mottling .... *Dermatobranchus* sp. 4 (214)
199A Cup-like sheath surrounding rhinophores (Fig. 7h) 200
199B Rhinophores sheath absent 214
200A Gills or cerata divided into compound structures 201
200B Gills or cerata undivided, papillae or tubercles may be present (Fig. 14) 210
201A Papillae bifid, animal purple with red rhinophores .... *Marianina rosea* (203)
201B Papillae more highly divided 202
202A Rhinophores sheaths with elongate extensions 203
202B Rhinophore sheaths without extensions, may have simple tubercles 204
203A Papillae at base of gills club-shaped .... *Bornella adamsi* (195)
203B Papillae at base of gills paddle-shaped .... *Bornella anguilla* (196)
204A Two pairs of fleshy lobes bear gills .... *Notobryon wardi* (197)
204B Fleshy lobes absent, many pairs of gills present 205
205A Paired blue spots present on notum .... *Marionia* sp. 2 (208)
205B Blue spots absent 206
206A Notum with ridges or many wart-like tubercles 207
206B Notum smooth 209
207A Notum with network of ridges .... *Marionia* sp. 1 (207)
207B Notum with wart-like tubercles 208
208A Body brown with opaque white tubercles .... *Marioniopsis cyanobranchiata* (209)
208B Body uniformly white, yellow, or orange .... *Tritonia nilsodhneri* (204)
209A Velar appendages elongate, bifid .... *Tritonia* sp. 1 (205)
209B Velar appendages short, tubercular .... *Tritonia* sp. 2 (206)
210A Head expanded into hood (Fig. 8c) 211
210B Head not expanded 212
211A Body pure white, rhinophore sheath with appendage .... *Melibe* sp. (199)
211B Body pink to red-orange, rhinophore sheath without appendage .... *Melibe rosea* (198)
212A Cerata without red, brown, or black spots .... *Doto rosea* (202)
212B Cerata with red, brown, or black spots 213
213A Ceratal tubercles elongate .... *Doto pinnatifida* (201)
213B Ceratal tubercles short (Fig. 14a) .... *Doto coronata* (200)
214A Convoluted crest present between rhinophores 215
214B Crest absent between rhinophores 217
215A Cerata yellow, blue, or orange, without digestive gland ducts .... *Bonisa nakaza* (218)
215B Cerata red brown (bluish epidermal pigment may be present) with digestive gland 216
216A Head rounded in front when viewed from above .... *Janolus longidentatus* (217)
216B Head conical in front when viewed from above .... *Janolus capensis* (216)
217A Cerata arranged in clusters elevated from notum (Fig. 12a) 218

217A	Cerata arranged in clusters elevated from notum (Fig. 12a)	218
217B	Ceratal clusters not elevated from notum (Fig. 12b)	224
218A	Rhinophores ornamented (Fig. 7d, e, f, g, i)	219
218B	Rhinophores smooth (Fig. 7c)	222
219A	Rhinophores annulate (Fig. 7e)	220
219B	Rhinophores papillate or perfoliate (Fig. 7d, f)	221
220A	Body purple with red cerata	*Flabellina funeka* (219)
220B	Body white with brown cerata and opaque white spots	*Amanda armata* (257)
221A	Rhinophores perfoliate (Fig. 7d)	*Flabellina* sp. 3 (223)
221B	Rhinophores papillate (Fig. 7f)	*Coryphellina* sp. (224)
222A	Body deep blue and silver	*Glaucus atlanticus* (264)
222B	Color not as above	223
223A	Mid-dorsal pink-purple line present	*Phyllodesmium serratum* (255)
223B	Mid-dorsal line absent	*Cuthona* sp. 3 (235)
224A	Rhinophores smooth (Fig. 7c)	225
224B	Rhinophores ornamented (Fig. 7d, e, f, g, i)	250
225A	Anterior corners of foot rounded (Fig. 13c)	226
225B	Anterior corners of foot angular or tentacular (Fig. 13a, b, d)	241
226A	Single ceras per ceratal row	227
226B	Two or more cerata per ceratal row	228
227A	Oral tentacles well developed	*Tergipes tergipes* (238)
227B	Oral tentacles flat, rounded, forming oral veil	*Embletonia gracilis* (228)
228A	Posterior edge of cerata with sail-like appendage (Fig. 14c)	*Fiona pinnata* (243)
228B	Cerata without appendage (Fig. 14a, b, d)	229
229A	Body translucent or opaque white, cerata variously colored	230
229B	Body with red, purple, orange, or brown pigment	236
230A	Cerata tuberculate	*Phyllodesmium hyalinum* (256)
230B	Cerata smooth	231
231A	Cerata numerous with blue pigment and yellow bands	*Cuthona kanga* (231)
231B	Cerata few, blue pigment absent	232
232A	Cerata largely red or brown	233
232B	Cerata largely opaque white	234
233A	Cerata brown, with subapical yellow spot	*Cuthona* sp. 5 (237)
233B	Cerata red-orange with white apex	*Catriona casha* (239)
234A	Cerata with subapical orange band	*Catriona* sp. (241)
234B	Cerata without orange band	235
235A	Cerata congested, completely covering notum	*Catriona columbiana* (240)
235B	Cerata scattered, notum visible	*Cuthona* sp. 1 (233)
236A	Body with purple pigment, cerata red	*Cuthona* sp. 4 (236)
236B	Body with orange or brown pigment	237
237A	Body with orange pigment	238
237B	Body with brown pigment	240
238A	Distinct gap present between rhinophores and first cerata	*Cuthona* sp. 2 (234)
238B	Cerata beginning immediately behind rhinophores	239
239A	Cerata inflated (Fig. 14b)	*Eubranchus* sp. 1 (225)
239B	Cerata cylindrical (Fig. 14d)	*Phestilla melanobrachia* (242)
240A	Rhinophores and oral tentacles without brown ring	*Eubranchus* sp. 2 (226)
240B	Rhinophores and oral tentacles with brown ring on outer half	*Eubranchus* sp. 3 (227)
241A	Body yellow or orange, cerata variably colored	242
241B	Body translucent white, often with pattern of orange pigment	244
242A	Opaque white lines present on head	*Caloria indica* (248)
242B	Opaque white lines absent on head	243
243A	Body yellow with orange pigment on head	*Cuthona ornata* (230)
243B	Body entirely orange	*Cuthona speciosa* (229)
244A	Foot corners short, slightly angular (Fig. 13d)	245
244B	Foot corners long, tentacular (Fig. 13b)	246

245A Foot broad, cerata congested .... *Aeolidiella indica* (265)

245B Foot narrow, ceratal groups well separated .... *Cratena capensis* (258)

246A Cerata opaque white over entire surface .... *Caloria* sp. 3 (251)

246B Cerata with reddish digestive gland 247

247A Notum with opaque white circles outlined in orange .... *Cratena simba* (259)

247A Notum with opaque white circles outlined in orange .... *Cratena simba* (259)

247B Notum without opaque white circles, pair of orange spots on head 248

248A Cerata with subapical black spot .... *Cratena* sp. 1 (260)

248B Cerata without subapical black spot 249

249A Cerata tipped with white .... *Cratena* sp. 2 (261)

249B Cerata tipped with orange or yellow .... *Cratena* sp. 4 (263)

250A Rhinophores papillate (Fig. 7f) 251

250B Rhinophores not papillate 254

251A Foot corners rounded 252

251B Foot corners tentacular 253

252A Notum and cerata with brown vein-like markings .... *Baeolidia palythoae* (268)

252B Notum and cerata with opaque white .... *Berghia chaka* (267)

253A Rhinophores and oral tentacles with opaque white pigment .... *Flabellina* sp. 1 (221)

253B Rhinophores and oral tentacles with red-orange pigment .... *Moridilla brockii* (247)

254A Rhinophores with two to four swellings or annulations (Fig. 7i) 255

254B Rhinophores with six or more annulations, rugose or perfoliate 258

255A Foot corners long, tentacular (Fig. 13b) 256

255B Foot corners short, angular, or rounded (Fig. 13a, c, d) 257

256A Cerata papillate .... *Favorinus japonicus* (252)

256B Cerata smooth .... *Favorinus ghanensis* (253)

257A Ceratal rows congested .... *Aeolidiella alba* (266)

257B Ceratal groups well separated .... *Cratena* sp. 3 (262)

258A Rhinophores perfoliate (Fig. 7d) 259

258B Rhinophores rugose or annulate (Fig. 7e, g) 261

259A Apex of cerata opaque white with subapical spot .... *Caloria* sp. 1 (249)

259B Apex of cerata without subapical spot 260

260A Body with scattered opaque white spots .... *Pruvotfolia pselliotes* (246)

260B Body translucent white throughout, apex of cerata yellow orange .... *Caloria* sp. 2 (250)

261A Rhinophores annulate with distinct lamellae (Fig. 7e) 262

261B Rhinophores rugose, without distinct lamellae (Fig. 7g) 264

262A Foot corners rounded .... *Cuthona annulata* (232)

262B Foot corners tentacular 263

263A Head and notum with orange markings .... *Facelina* sp. (245)

263B Head and notum without orange markings .... *Facelina olivacea* (244)

264A Cerata with yellow and blue or purple bands .... *Godiva quadricolor* (254)

264B Cerata without yellow or blue bands 265

265A Head and cerata with opaque white lines .... *Flabellina capensis* (220)

265B Head and cerata with opaque white spots .... *Flabellina* sp. 2 (222)

Species Accounts

Order Cephalaspidea

Family Acteonidae

1. ***Rictaxis albus*** (Sowerby, 1873)

Taxonomy: This species was originally described as a species of *Acteon* and was later transferred to the genus *Pseudacteon*, on the basis of its radular morphology. Recently, Kilburn and Rippey (1982) correctly placed this species within *Rictaxis*.

Natural History: Living specimens of this species have been collected on mixed sandy and rocky reef in about 10 m of water. This species, like other members of the Acteonidae, probably feeds on polychaete annelid worms.

Occurrence and Distribution: Living specimens have been collected only from Algoa Bay, although shells have been found on beaches from False Bay to Natal. This species appears to be endemic to southern Africa.

2. ***Japonacteon* sp.**

Taxonomy: The elevated shell (Fig. 4a) with black markings readily differentiates this species from all other opisthobranchs dealt with in the study. It differs externally and internally from all other members of the Acteonidae recorded from southern Africa. It is the only species with a white shell with dark brown checks. This undescribed species is regarded as belonging to *Japonacteon* by virtue of its unique radular teeth. The only other species known to belong to this genus is *Japonacteon nipponensis*.

Natural History: *Japonacteon* sp. is found in stable sandy flats in water about 3 m deep. Although no observations have been made of this species when feeding, it probably feeds on annelid worms.

Occurrence and Distribution: *Japonacteon* sp. has been collected only in South Africa. A single specimen of this undescribed species was found near Vetchies Pier on the north side of Durban Harbor.

Family Hydatinidae

3. ***Micromelo undata*** (Bruguière, 1792)

Taxonomy: The shell of *Micromelo undata* is unique in that there are three spiral brown lines with several irregular, inter-connecting lines of the same color. There is no obvious shell sculpture as in the closely related *Bullina scabra*. The animal is a beautiful blue-green with yellow lines and white spots. Previous records of this species from South Africa erroneously identified it as *Bullina ziczac*.

Natural History: *Micromelo undata* is common in rocky pools in the lower intertidal zone. It feeds on polychaete worms.

Occurrence and Distribution: Numerous specimens have been found intertidally along the Natal coast from Park Rynie to Kosi Bay. *Micromelo undata* is circumtropical, being widely distributed throughout the Indo-Pacific and Caribbean Sea.

4. ***Hydatina physis*** (Linnaeus, 1758)

Taxonomy: The shell has numerous spiral brown or black lines. The animal is dark pink to red with an opaque white or blue marginal line around the foot.

Natural History: This species is a highly specialized predator, feeding exclusively on polychaete worms of the family Cirratulidae. *Hydatina physis* inhabits sand flats and rock pools. Its egg mass is highly convoluted and is attached to the sand by a mucus thread.

Occurrence and Distribution: *Hydatina physis* has been found in southern Africa from Port Elizabeth northwards to Mozambique. This species is circumtropical and is known from all warm oceans of the world.

5. ***Hydatina amplustre*** (Linnaeus, 1758)

Taxonomy: The alternating pink and white spiral bands, separated by narrow black lines readily differentiate this shell from all other species. The animal is a uniform translucent white.

Natural History: Like the preceding species, *Hydatina amplustre* is a specialized predator on cirratulid polychaete worms. It inhabits areas of mixed sand and rock. At Salt Rock, Natal, it has been found together with *H. physis*.

Occurrence and Distribution: *Hydatina amplustre* is generally uncommon, but has been found in southern Africa from Durban to Mozambique. Macnae (1962*b*) reported that this species is common at Inhaca. *Hydatina amplustre* is found throughout the tropical Indo-Pacific, but is absent from the Atlantic.

Family Scaphandridae

6. ***Cylichna tubulosa*** Gould, 1859

Taxonomy: The shell is elongate and cylindrical (Fig. 4b) with a brownish periostracum covering the white shell. The living animal is uniformly translucent white.

Natural History: *Cylichna tubulosa* inhabits fine mixed sand and mud bottoms, generally well below the tide mark, in 10-300 m of water. *Cylichna tubulosa* feeds on foraminiferans; fragments of these calcareous protozoans have been found in the gizzards of specimens.

Occurrence and Distribution: *Cylichna tubulosa* has been found from Angola to Durban and appears to be endemic to southern Africa.

Family Retusidae

7. ***Retusa truncatula*** (Bruguière, 1792)

Taxonomy: *Retusa truncatula* is distinguished by its minute shell with a flattened spire. The animal is white and is generally less than 4 mm in length. The shell has minute longitudinal striations on its surface.

Natural History: *Retusa truncatula* lacks a radula. It feeds on foraminiferans by crushing their calcareous shells with its three gizzard plates. This species is found in shallow estuaries where it burrows in mud or sand in tidal marshes.

Occurrence and Distribution: In South Africa, *Retusa truncatula* has been found from False Bay to Durban. It is found intertidally or just below the low water mark. It is very common in Knysna Lagoon, but may be easily overlooked because of its size. *Retusa truncatula* is also known from northern Europe, the Mediterranean, and the Canary Islands.

Family Philinidae

8. ***Philine aperta*** (Linnaeus, 1767)

Taxonomy: The large (up to 100 mm) white or cream body with a blunt tail (Fig. 6b) readily distinguishes this species. The internal shell from dead specimens is often found on beaches.

Natural History: *Philine aperta* is a predatory gastropod, feeding on a variety of organisms, including foraminiferans, small snails, and bivalves. It is probably one of the most common snails in South Africa in sandy habitats. Frequently, the oblong egg masses are found attached to the sandy bottom by an elongate mucus thread.

Occurrence and Distribution: *Philine aperta* inhabits sandy areas from the low tide mark to several hundred meters of depth. It is commonly found in tidal swimming pools which have a sandy bottom. It appears to be common along most of the southern African coast where it is known from Saldanha Bay to Mozambique (Macnae 1962*b*). This species is very widely distributed along the eastern Atlantic, from northern Europe to South Africa, and into the Indian and Pacific oceans. In tropical regions it is generally found in much deeper water.

Family Aglajidae

9. ***Philinopsis cyanea*** (Martens, 1879)

Taxonomy: *Philinopsis cyanea* is perhaps one of the most variably colored of all opisthobranchs (Eliot 1903; Macnae 1962*b*; Rudman 1972). The body ranges in color from tan to blue-black.

Natural History: *Philinopsis cyanea* lacks a radula and engulfs its prey whole by suction with its muscular pharynx. Generally, this species feeds on other opisthobranchs and commonly eats the bubble snail, *Haminoea natalensis*. It generally burrows in sand or mud and tracks its prey by using its chemoreceptors to follow the mucus trail of its victim. The egg mass is virtually identical to that of *Philine aperta*.

Occurrence and Distribution: Macnae (1962*b*) found this species at Inhaca in Mozambique and recently specimens have been found in the tidal swimming pool on the Natal North Coast at Salt Rock. This species is widely distributed throughout the Indo-Pacific.

10. ***Philinopsis capensis*** (Bergh, 1907)

Taxonomy: Until recently, this species was thought to be synonymous with the preceding species. The two species are very similar in coloration and general appearance. However, they differ in details of the male reproductive system.

Natural History: *Philinopsis capensis* is similar to *P. cyanea* in its mode of feeding and choice of prey. Although Bergh (1907) noted that it feeds on worms, it also appears to feed upon the opisthobranchs *Haminoea alfredensis* and small specimens of *Akera soluta*. This species readily burrows in fine sand.

Occurrence and Distribution: Specimens have been found in tidal swimming pools and in estuaries from False Bay to East London. It appears to be endemic to the warmer waters of the Cape Province.

11. ***Chelidonura hirundinina*** (Quoy and Gaimard, 1824)

Taxonomy: *Chelindonura hirundinina* can be distinguished by brown to black coloration with a "T" shaped marking on its head (Fig 6a). Until recently, the form with brown coloration was regarded as a distinct species, *C. philinopsis*. At Umgazana, Transkei, both forms are found together with specimens of intermediate coloration. Black and brown individuals freely reproduce with each other.

Natural History: *Chelidonura hirundinina* has been observed in the field feeding upon small flat worms. It everts its pharynx and sucks up its prey, as does *Philinopsis*. *Chelidonura hirundinina*, unlike *Philinopsis*, crawls on the surface of sand rather than burrowing and is often found in rock pools. Round globular egg masses are deposited on rock surfaces.

Occurrence and Distribution: Previously, *Chelidonura hirundinina* was known in southern Africa only from Inhaca. It extends at least as far south as Umgazana, where it can be very common in rock pools. This species is circumtropical and is known from the Indian, Pacific and Atlantic oceans.

12. ***Chelidonura fulvipunctata*** Baba, 1938

Taxonomy: *Chelidonura fulvipunctata* is distinguishable from the preceding species by the white "W" shaped marking on its head shield.

Natural History: Little or nothing is known about the biology of this species; it was found under rocks in 2 m of water.

Occurrence and Distribution: The only South African record of this species is of a single specimen collected under rocks at the mouth of the Knysna Lagoon. *Chelidonura fulvipunctata* has also been found in Hawaii, Guam, Australia, and Turkey, suggesting that it is a circumtropical species.

13. ***Chelidonura* sp.**

Taxonomy: This species is distinct from all other members of the genus and appears to be undescribed. It is the only *Chelidonura* with uniformly white coloration.

Natural History: Specimens have been found underneath rocks in sandy areas. The prey of *Chelidonura* sp. is unknown.

Occurrence and Distribution: This species is known only from the intertidal region of Gonubie, north of East London.

Family Gastropteridae

14. ***Gastropteron flavobrunneum*** Gosliner, 1984

Taxonomy: This species was recently described and is distinguished by its yellow body with brown spots. The Gastropteridae are similar in appearance to the Aglajidae but possess a radula and have a more concentrated nervous system.

Natural History: *Gastropteron flavobrunneum* inhabits rocky reefs from a depth of 15 to 20 m. It is capable of swimming by flapping its parapodia.

Occurrence and Distribution: Only three specimens of this species have been found. They were collected from the Atlantic coast of the Cape Peninsula, at Bakoven and Llandudno. *Gastropteron flavobrunneum* appears to be endemic to the Cape Province.

15. ***Gastropteron alboaurantium*** Gosliner, 1984

Taxonomy: *Gastropteron alboaurantium* is distinguishable by its pinkish color, with orange pigment on the visceral hump and on the tip of the siphon. Like *Gastropteron flavobrunneum*, it was also recently described.

Natural History: This species, like the preceding one, inhabits rocky reefs at about 20 m depth. It also swims by flapping its parapodia.

Occurrence and Distribution: This species has been found at Bakoven together with a specimen of *Gastropteron flavobrunneum*. It appears to be endemic to southern Africa.

Family Haminoeidae

16. ***Haminoea alfredensis*** Bartsch, 1915

Taxonomy: *Haminoea alfredensis* (Fig. 4c) has a broad body that is yellowish green with yellow spots. Internally, it is very different from the other described South African species.

Natural History: Like other members of this genus, *Haminoea alfredensis* is a grazing herbivore. It is most commonly found in shallow estuaries, where it is exceedingly abundant on eel grass, from which it scrapes diatoms. Occasionally, it occurs in high rock pools or in tidal swimming pools, such as in False Bay, where it feeds on filamentous green algae.

Occurrence and Distribution: *Haminoea alfredensis* is found on both sides of the Cape Peninsula and eastward to north of East London. It is endemic to temperate South African waters.

17. ***Haminoea natalensis*** (Krauss, 1848)
Synonyms: *Haminoea petersi* Martens, 1879
Haminoea gracilis Sowerby, 1897
Haminoea subcylindrica Sowerby, 1897

Taxonomy: The four species of *Haminoea* listed above were regarded as distinct by Macnae (1962*b*), based on slight difference in the shell and radula. However, they seem to fall within the range of variation of a single species. *H. natalensis* can be distinguished from *H. alfredensis* by its coloration and by the narrow aperture of the shell.

Natural History: This species is found in rock pools and tidal swimming pools where it feeds on filamentous green algae.

Occurrence and Distribution: *Haminoea natalensis* is found in shallow water from Port Alfred to Mozambique. It may be synonymous with other Indo-Pacific species although the taxonomy of the group is poorly understood and meaningful comparisons cannot be made at this time.

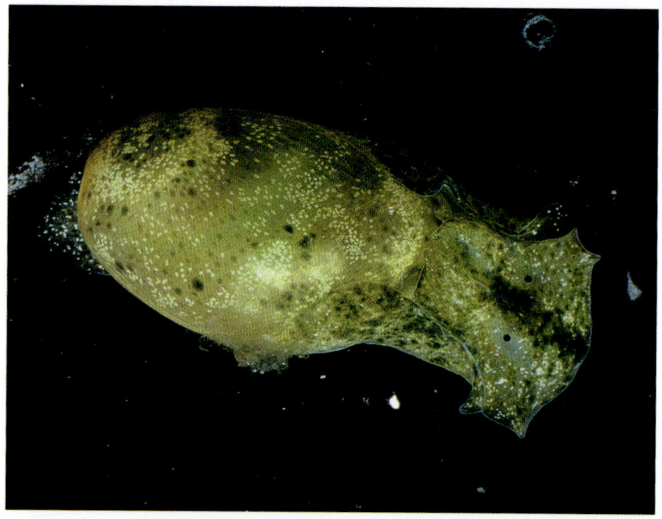

18. ***Smaragdinella calyculata*** (Broderip and Sowerby, 1829)

Taxonomy: The broad body, with parapodia extending the entire length of the animal, distinguishes *Smaragdinella* from species of the closely related genus *Haminoea*.

Natural History: *Smaragdinella* is common in high rock pools but is rarely seen. It is found together with *Haminoea natalensis*. The animals are well camouflaged, blending in well with the algal film growing on the rocks. They appear to be nocturnal.

Occurrence and Distribution: Specimens have been collected only from Jesser Point, Sodwana Bay National Park. This constitutes the first record of the species from southern Africa. *Smaragdinella calyculata* is widely distributed in the Indo-Pacific tropics as far east as Hawaii.

19. ***Phaneropthalmus smaragdinus*** (Rüppell and Leuckart, 1831)

Taxonomy: This species varies in color from white to dark green. It appears to lack an external shell, but if the parapodia are opened, a small opaque white shell plate is visible on the posterior end of the animal.

Natural History: *Phaneropthalmus smaragdinus* has been collected in intertidal rock pools. It presumably feeds upon filamentous green algae and also appears to be nocturnal.

Occurrence and Distribution: Numerous specimens were found at Jesser Point, Sodwana Bay National Park. This is the first record of the species from southern Africa. This species is widespread throughout the Indo-Pacific as far east as Easter Island.

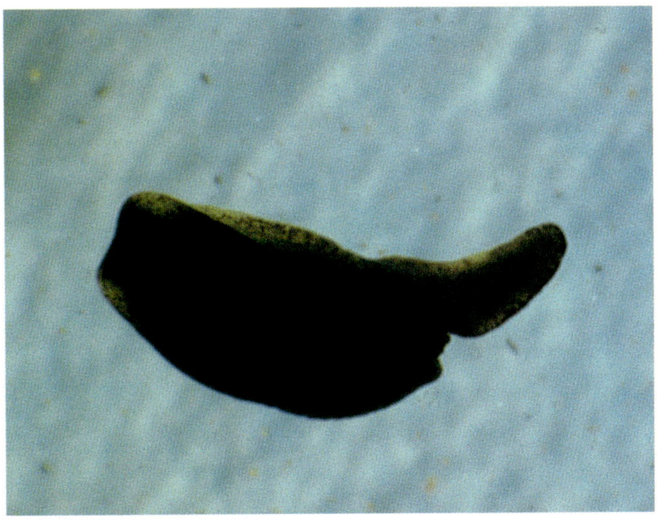

Family Runcinidae

20. ***Metaruncina* sp.**

Taxonomy: This tiny, undescribed opisthobranch reaches a maximum length of 3 mm and resembles a small black flatworm. This species is distinguished from other runcinids by the shape of its internal shell, lack of gizzard plates and single, folded gill. The only other known species of the genus is *M. setoensis* from Japan, which has a different radular form and possesses a gizzard.

Natural History: *Metaruncina* sp. is found among the roots and stipes of the siphonaceous green alga *Caulerpa racemosa* in the lower intertidal realm.

Occurrence and Distribution: Several specimens have been collected in the vicinity of Sodwana Bay National Park.

Family Bullidae

21. ***Bulla ampulla*** (Linnaeus, 1758)

Taxonomy: *Bulla ampulla* differs from *Haminoea* by its thicker, opaque, rather than translucent shell, which has a mottled pattern.

Natural History: *Bulla ampulla* and other members of the genus appear to be strict herbivores, although Macnae (1962*b*) stated that it feeds on other mollusks. *B. ampulla* is always found around soft silt or mud in which it is usually buried.

Occurrence and Distribution: Shells have been found along the coast of southern Africa from False Bay to Mozambique. Living animals have been found only from Port Elizabeth northwards. They are common in deep rock pools and tidal swimming pools, from Umgazana, Transkei to Salt Rock, Natal. *Bulla ampulla* is widespread throughout the Indo-Pacific.

Order Theocosomata

Family Cavoliniidae

22. ***Creseis acicula*** (Rang, 1828)

Taxonomy: *Creseis acicula* can be distinguished by its elongate, conical shell. Other shelled thecosome pteropods are common off the coast of southern Africa but have not been collected alive.

Natural History: *Creseis acicula* is one of the few opisthobranchs which spends its entire life drifting in the plankton. It is a filter feeder and strains microscopic organisms, such as diatoms, from the water.

Occurrence and Distribution: Specimens have been collected just below the water surface off Park Rynie, Natal. *Creseis acicula* is found throughout the subtropical and tropical oceans of the world.

Order Anaspidea

Family Akeridae

23. ***Akera soluta*** (Gmelin, 1791)

Taxonomy: This species has a thin, brown bulloid shell. Its head-shield is narrow and elongate.

Natural History: *Akera soluta* is a grazing herbivore. It is generally found in estuaries or tidal swimming pools where it burrows through sandy substrate. When disturbed, it is able to swim by flapping its parapodia. Its egg mass consists of a convoluted cord of mucus.

Occurrence and Distribution: *Akera soluta* has been found intertidally and in shallow waters from Knysna Lagoon to Mozambique. It is exceedingly common in tidal swimming pools along the Natal coast. This species has been found throughout the Indian Ocean, along the Pacific coast of eastern Australia and as far east as Enewetak Atoll.

Family Aplysiidae

24. ***Aplysia parvula*** Mörch, 1863

Taxonomy: *Aplysia parvula* (Fig. 5c) is exceedingly variable in its coloration, particularly in specimens from the Cape Peninsula. In almost all cases there are clusters of opaque white spots on the outer surface of the parapodia, but, in a few instances, they are entirely absent. It is the smallest of the *Aplysia* species, reaching a maximum length of 50 or 60 mm.

Natural History: Like other sea hares, *Aplysia parvula* is a specialized herbivore and feeds largely on red algae. It is common in intertidal and shallow subtidal habitats, to about 5 m in depth. Its egg mass, which is commonly found under rocks, is a tangled network of sticky mucus strings.

Occurrence and Distribution: *Aplysia parvula* has been found along the South African coast, from the Atlantic side of the Cape Peninsula to northern Natal. *Aplysia parvula* is circumtropical.

25. ***Aplysia oculifera*** Adams and Reeve, 1850

Taxonomy: In the South African literature, *Aplysia oculifera* was originally united with *A. dactylomela* as a single species. It is clearly distinguishable by having smaller black rings on the body and a far more graceful appearance. It reaches approximately 150 mm in length.

Natural History: This species is found in shallow estuaries and embayments. It is an herbivore that is commonly nocturnal.

Occurrence and Distribution: *Aplysia oculifera* has recently been found from Dale Brook in False Bay and the Keurbooms River estuary at Plettenberg Bay to the Natal north coast. Earlier records must be discounted owing to the confusion of the species with *A. dactylomela*. *Aplysia oculifera* is widely distributed throughout the Indo-Pacific as far east as the Hawaiian Islands.

26. ***Aplysia dactylomela*** Rang, 1828

Taxonomy: *Aplysia dactylomela* is characterized by its green or brown ground color with large black circles. This species is large and may reach 400 mm in length.

Natural History: *Aplysia dactylomela* is common on tidal platforms of mixed sand and rock. Like many sea hares, *A. dactylomela* is largely nocturnal. Animals are often seen actively crawling in tide pools at night, but during the day the animals remain inactive in crevices and under rocks.

Occurrence and Distribution: *Aplysia dactylomela* is common intertidally and in the shallow subtidal zone from Port Elizabeth to Mozambique. It is particularly abundant along the Transkei coast. Like many of the sea hares, *Aplysia dactylomela* has a circumtropical distribution.

27. ***Aplysia juliana*** Quoy and Gaimard, 1832

Taxonomy: Although this species is highly variable in its coloration, it can be readily distinguished by the sucker on the posterior end of the foot.

Natural History: *Aplysia juliana*, like most other anaspideans, is nocturnal. Specimens at Salt Rock remained hidden within crevices during the day. At night, the animals were actively crawling near the surface of the water, where they were observed feeding on *Ulva*, copulating, and depositing egg masses.

Occurrence and Distribution: *Aplysia juliana* has been found in southern Africa from Bloubergstrand, on the Atlantic coast of the Cape Peninsula, to Mozambique. This species is circumtropical.

28. ***Dolabella auricularia*** (Solander, 1786)

Taxonomy: There appears to be only a single species in this genus. It is recognizable by the flattened disc on the posterior end of the animal. Despite its large size (over 400 mm in length), it is often cryptically colored when on rocks and sand and is therefore rarely seen.

Natural History: *Dolabella auricularia* actively crawls around eel grass beds, reef flats, or in tidal swimming pools during the night or on overcast days. During the day it is found under rocks or buried in sand or mud.

Occurrence and Distribution: *Dolabella auricularia* has rarely been recorded from the South African coast. Macnae (1962*b*) stated that it was common on reef flats in Inhaca and that single specimens were known from the Umtata River mouth in Transkei and from Durban. Its apparent rarity can be attributed to the fact that it is cryptically colored and nocturnal. Specimens have recently been found at the Keurbooms River Estuary at Plettenberg Bay. The species is very abundant on reef flats at Umgazana, Transkei, as well as along the entire Natal coast. *Dolabella auricularia* is widely distributed throughout the Indo-Pacific.

Family Notarchidae

29. ***Bursatella leachi*** (Blainville, 1817)
Synonym: *Barnardaclesia cirrhifera* (Quoy and Gaimard, 1832)

Taxonomy: *Bursatella leachi* has been subdivided into several subspecies, two of which occur in South African waters. *B. leachi africana* (A) has shortened papillae and no rings on the body surface. *B. leachi leachi* (B) possesses blue circles on its body surface. *Barnardaclesia* was previously regarded as a distinct genus and species, but appears to be identical with *Bursatella leachia leachi* (present study).

Natural History: Both sub-species are commonly found in estuaries and tidal swimming pools. Their population numbers fluctuate sporadically. Thousands of individuals can be present in a small area of a lagoon, in very dense aggregations, but may be entirely absent a few weeks later.

Occurrence and Distribution: *Bursatella leachi africana* has been found from Langebaan Lagoon and Sandvlei, which empties into False Bay, to East London. *B. leachi leachi* has been found from the Natal coast to Mozambique. *Bursatella leachi* is considered a circumtropical species although its sub-species are geographically isolated. *B. leachi africana* is endemic to the Cape Province while *B. leachi leachi* is widely distributed throughout the Indo-Pacific.

(A)

(B)

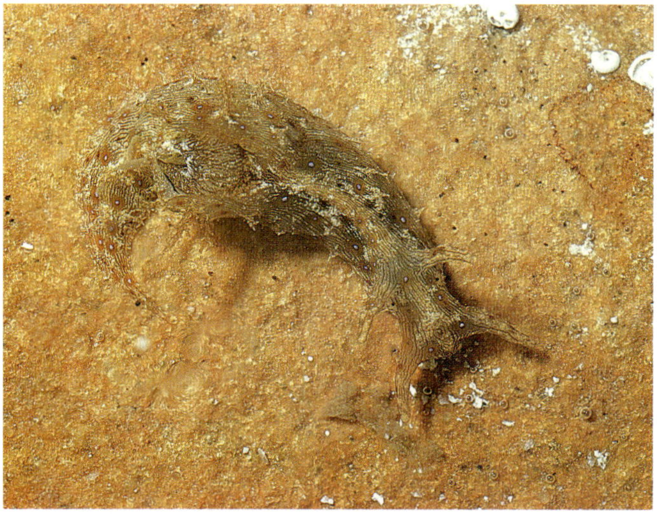

30. ***Stylocheilus longicauda*** (Quoy and Gaimard, 1824)

Taxonomy: *Stylocheilus longicauda* is distinguishable by its simple papillae, longitudinal lines and blue spots. It is generally much smaller and more slender than *Bursatella leachi.*

Natural History: This species is found in sheltered pools and estuaries in the same habitat as *Bursatella leachi*, where it feeds on a variety of algae.

Occurrence and Distribution: Macnae (1962*b)* recorded this species only from Inhaca, but recent investigations have found specimens as far south as Umgazana, Transkei. This species is circumtropical.

Order Sacoglossa

Family Cylindrobullidae

31. ***Ascobulla fischeri*** (Adams and Angas, 1864)
Synonyms: *Cylindrobulla pusilla* Nevill and Nevill, 1869
Cylindrobulla sculpta Nevill and Nevill, 1869

Taxonomy: Marcus (1972) has shown that although *Cylindrobulla* is externally indistinguishable from *Ascobulla*, the two genera have very different radular teeth. The radula of the South African species is clearly that of *Ascobulla*. Although two other species of "*Cylindrobulla*" have been recorded from South Africa (Sowerby 1894, 1897) their taxonomy is based only upon conchological differences. The present material closely resembles *Ascobulla fischeri*, which is an older name.

Natural History: *Ascobulla fischeri* occurs on the green algae, *Caulerpa racemosa* and *C. filiformis* in shallow rock pools and channels.

Occurrence and Distribution: *Ascobulla fischeri* has been found from Strandfontein in False Bay to Ramsgate on the Natal South Coast. *Ascobulla fischeri* is most probably widely distributed throughout the Indo-Pacific.

32. ***Volvatella laguncula*** Sowerby, 1894
Synonym: *Cylindrobulla turtoni Bartsch, 1915*

Taxonomy: *Volvatella laguncula* was previously known only from two empty shells (Fig. 4e) from the type localities. Thompson (1979) has recently reviewed the taxonomy of the genus based on the study of living material of *V. laguncula* from False Bay.

Natural History: Thompson has shown that *Volvatella laguncula* feeds exclusively on the roots of *Caulerpa filiformis*. Although few species of this genus have been studied thoroughly, the burrowing habit of this species appears to be the primitive condition. In Hawaii, I found *Volvatella fragilis* crawling only on the blades of *Caulerpa* spp. rather than burrowing.

Occurrence and Distribution: *Volvatella laguncula* has been found intertidally from Strandfontein, in False Bay, and Port Elizabeth. It appears to be endemic to South Africa. A single subtidal specimen was found at a depth of 8 m at Windmill Beach, False Bay.

Family Juliidae

33. ***Berthelinia schlumbergeri*** Dautzenberg, 1895

Taxonomy: The taxonomy of the bivalved gastropods is controversial. Several workers have subdivided them into four genera. Kay (1968) stated that this subdivision is unnecessary and regarded all the bivalved gastropods, except *Julia*, as belonging to *Berthelinia*. Many of the species described in the genus *Tamaovalva* differ only slightly in the shape of their shells and probably represent variations of a single species. For this reason, the South African specimens are regarded as *Berthelinia schlumbergeri*, the oldest species name of the group.

Natural History: *Berthelinia schlumbergeri* is found in the rocky intertidal on the green alga, *Caulerpa racemosa*.

Occurrence and Distribution: Two specimens of *Berthelinia schlumbergeri* were recently found at Mbibi, 20 km north of Sodwana Bay National Park. This represents the first record of a bivalved gastropod from southern Africa. *Berthelinia schlumbergeri* is probably distributed throughout the Indo-Pacific.

34. ***Julia zebra*** Kawaguti, 1981

Taxonomy: *Julia* differs from *Berthelinia* by the convex shape of its shell and by its darker coloration. In *Julia*, the posterior end of the shell is deeply indented and possesses numerous dark brown lines.

Natural History: Like *Berthelinia*, this species is associated with the alga, *Caulerpa racemosa*. It is found at the lower fringes of the rocky intertidal region.

Occurrence and Distribution: A single specimen was collected at Mbibi, 20 km north of Sodwana Bay National Park. This represents the first record of the species from southern Africa. This species is known from Japan and Easter Island.

Family Oxynoidae

35. ***Lobiger souverbiei*** Fischer, 1856

Taxonomy: Baba (1974) has shown that *Lobiger souverbiei* is widely distributed and apparently differs from *L. viridis* Pease, 1863 and *L. serradifalci* (Cabara, 1840) although the anatomical differences are few. *Lobiger* (Fig. 51) differs from other genera in having four parapodial lobes and a shell with a straight inner margin. The parapodia can be shed and are often damaged, as seen in the specimen on the right.

Natural History: *Lobiger souverbiei*, like the four preceding species, *Oxynoe virdis* and *Elysia* sp., is a member of the caulerpan microfauna and has been found on *Caulerpa racemosa*.

Occurrence and Distribution: Two specimens of *Lobiger souverbiei*, found at Sodwana Bay National Park, represent the first records of this species from southern Africa. *Lobiger souverbiei* is circumtropical in its distribution.

36. ***Oxynoe viridis*** (Pease, 1861)
Synonym: *Oxynoe natalensis* Smith, 1903

Taxonomy: *Oxynoe viridis* is readily distinguishable from *Oxynoe* sp. on the basis of its coloration and shell shape. The closely related species, *Lophopleurella capensis*, is known only from the original description of one specimen from off Simonstown in False Bay. It may be confused with *Oxynoe*, but can be distinguished by its shell, which resembles that of *Lobiger*.

Natural History: *Oxynoe viridis* has been found on *Caulerpa racemosa* in rocky, tropical habitats along the South African coast.

Occurrence and Distribution: *Oxynoe viridis* (as *O. natalensis*) was previously known only from dry shells collected near Durban. Numerous living specimens have recently been collected from the Natal South Coast at Margate and Ramsgate and from Gonubie near East London. *Oxynoe viridis* is widely distributed throughout the Indo-Pacific.

37. ***Oxynoe* sp.**

Taxonomy: This species (Fig. 5b) represents an undescribed form that is clearly distinct from all other *Oxynoe* species in its coloration and internal anatomy.

Natural History: *Oxynoe* sp. is found crawling on sand flats at low tide. All other known species in the genus are associated with the green algae, *Caulerpa* spp.

Occurrence and Distribution: This species is known only from Knysna Lagoon, where it is sporadically common, and from False Bay.

Family Elysiidae

38. ***Elysia viridis*** (Montagu, 1804)
Synonym: *Elysia punctata* var. *rubropunctata* Macnae, 1954 non Kelaart, 1857

Taxonomy: Macnae (1954*b*) originally attributed this species (Fig. 5d) to South African waters under the name *Elysia punctata* var. *rubropunctata*. Macnae's species is very different from Kelaart's and closely resembles the common European species, *E. viridis*. As no morphological difference could be determined between the European and South African forms, they are considered the same species.

Natural History: *Elysia viridis* feeds upon several species of siphonaceous green algae, especially *Codium fragile* and *C. stephensiae*. It has also been found on *Bryopsis* and *Halimeda*. Specimens found on *Bryopsis* are much larger than those on other algae, reaching 30 mm in length. The egg mass is a flat, greenish coil, consisting of three or four whorls.

Occurrence and Distribution: *Elysia viridis* is one of the most common opisthobranchs along the South African coast, although it is seldom seen, owing to its ability to blend in with its algal prey. It has been found from Eland's Bay, on the Atlantic coast, to Natal. *Elysia viridis* is known from Northern Europe and the Mediterranean and may have been found in China.

39. ***Elysia halimedae*** Macnae, 1954

Taxonomy: *Elysia halimedae* is characterized by its stout body and short parapodia.

Natural History: *Elysia halimedae* is usually associated with the green alga, *Halimeda cuneata*, on which it feeds. Macnae (1954*b*) also found it feeding on *Codium* sp., a closely allied green alga.

Occurrence and Distribution: *Elysia halimedae* has been found in southern Africa from Port Alfred to Mozambique. Since the description of this species from South Africa, it has been reported from several localities throughout the Indo-Pacific, including Japan, Australia, Guam, and Hawaii. Previous records of the species, from Japan and Australia, have recently been shown to represent distinct species.

40. ***Elysia marginata*** (Pease, 1871)
Synonym: *Elysia ornata* (Pease, 1869) non Swainson, 1840

Taxonomy: Marcus (1980) has shown that *Elysia ornata* (Swainson, 1840) is distinct from *E. marginata* in the manner of the branching of the blood vessels from the heart. Macnae and Kalk's (1958) record of *E. ornata* from southern Africa must be regarded as being *E. marginata*.

Natural History: Specimens were found feeding on the green alga, *Codium* sp., in a tidal swimming pool and in rock pools.

Occurrence and Distribution: Macnae and Kalk (1958) recorded this species from Inhaca Island, Mozambique. Recently, several specimens have been found at Salt Rock, on the Natal North Coast, and at Sodwana Bay. It is widespread in the Indo-Pacific region.

41. ***Elysia virgata*** (Bergh, 1888)

Taxonomy: *Elysia virgata* is recognizable by its blue-green color, black longitudinal stripes and orange rhinophores.

Natural History: *Elysia virgata* is found on shallow reefs and has been found together with *Elysia moebii* and *E.* sp. 1.

Occurrence and Distribution: Several specimens were found in the vicinity of Sodwana Bay National Park. *Elysia virgata* was described by Bergh (1888) from a single specimen collected at Mauritius and had not been recorded again until it was discovered at Sodwana Bay.

42. ***Elysia moebii*** (Bergh, 1888)

Taxonomy: *Elysia moebii* is similar in body form to *E. virgata* but has a blue-green color with black and burnt orange parapodial bands spotted with opaque white.

Natural History: This species was found with *Elysia virgata* and *E.* sp. 1 on a shallow tropical reef.

Occurrence and Distribution: The single specimen was found at Mbibi, 20 km north of Sodwana Bay National Park. Like *Elysia virgata*, *Elysia moebii* was only known from a single specimen described from Mauritius in 1888 by Bergh. In addition to the South African material, a single specimen of this species was recently collected from Reunion Island by Michael Gosliner and Gary Williams.

43. ***Elysia vatae*** Risbec, 1928

Taxonomy: *Elysia vatae* is another one of the graceful, brightly colored species of *Elysia* found on tropical reefs. This species can be identified by its blue-grey color with black and yellow spots and red-orange tipped rhinophores.

Natural History: This species was found on dead coral reefs in less than 1 m of water. No direct association with its prey species was observed.

Occurrence and Distribution: The single specimen found at Adlam's Reef, Sodwana Bay National Park, represents the first record of *Elysia vatae* from the Indian Ocean. This species was described from New Caledonia and has recently been recorded from Guam.

44. ***Elysia livida*** Baba, 1955

Taxonomy: The black body, with orange, blue, and black lines on the parapodia and head, distinguishes *Elysia livida* from closely related members of the genus.

Natural History: This species is found in depths of less than 2 m on dead coral reefs.

Occurrence and Distribution: The three specimens collected around Sodwana Bay National Park represent the first record of *Elysia livida* from the Indian Ocean. *Elysia livida* was described from Japan and has recently been found in Guam. It is probably widespread throughout the Indo-Pacific and has recently been collected from Aldabra Atoll.

45. ***Elysia rufescens*** (Pease,1871)

Taxonomy: This species closely resembles the original description by Pease. Specimens from Hawaii identified by Kay (1980) appear to represent a distinct species. *Elysia rufescens* differs from all known members of the genus by its green and opaque white mottling and purple rhinophores and parapodial margin.

Natural History: *Elysia rufescens* is found on shallow, inshore reefs where it crawls over assemblages of algae.

Occurrence and Distribution: The single specimen was found at Mbibi, 20 km north of Sodwana Bay National Park. The species was originally described from Tahiti.

46. ***Elysia* sp. 1**

Taxonomy: This undescribed species can be distinguished by its green ground color with orange, black, and discontinuous blue parapodial markings. It is similar to *Elysia picta* from the Caribbean, but can be distinguished by its straight, rather than undulate parapodial lines.

Natural History: *Elysia* sp. 1 is found on shallow reefs and is the most common of the colorful tropical *Elysia* species in northern Natal.

Occurrence and Distribution: *Elysia* sp. 1 has been found in southern Africa only at Sodwana Bay National Park and vicinity. It has been recently collected from Aldabra and is likely more widespread in the Indo-Pacific.

47. ***Elysia* sp. 2**

Taxonomy: The bluish coloration, with opaque white parapodial lines and spots and red-orange rhinophores, distinguishes this species from all described members of the genus.

Natural History: This species was found at a depth of 2 m on a dead coral reef.

Occurrence and Distribution: The single specimen found was collected at Mbibi, 20 km north of Sodwana Bay National Park. It has also been collected from Aldabra Atoll.

48. ***Elysia* sp. 3**

Taxonomy: This species is undescribed. It most closely resembles *Elysia* sp. 1, but differs by its uninterrupted, undulating blue parapodial line. It also has two yellow parapodial bands, rather than the single orange band present in *E.* sp. 1.

Natural History: Like the preceding seven species, *Elysia* sp. 3 is found on shallow reef flats, in less than 2 m of water.

Occurrence and Distribution: A single specimen of this species was found at Adlam's Reef, Sodwana Bay National Park.

49. ***Elysia* sp. 4**

Taxonomy: This species appears to be undescribed. It differs from all South African species with parapodial papillae in having a patch of blue pigment on the head.

Natural History: *Elysia* sp. 4 is found on the green alga, *Caulerpa racemosa*, in the mid-intertidal zone.

Occurrence and Distribution: A single specimen of this species has been found at Sodwana Bay National Park.

50. ***Elysia* sp. 5**

Taxonomy: This species appears to be undescribed. Its light green color and its papillae on the parapodia and rhinophores are unique.

Natural History: Specimens were found in intertidal rock pools. This species appears to be nocturnal.

Occurrence and Distribution: Two specimens were collected at Jesser Point, Sodwana Bay National Park.

51. ***Elysia* sp. 6**

Taxonomy: *Elysia* sp. 6 is undescribed. It can be distinguished from those other members of the genus that have papillae on the parapodia by having brownish purple bands on the rhinophores and minute brown speckles on the head.

Natural History: This species was found together with the preceding species in intertidal rock pools, at night.

Occurrence and Distribution: The single specimen of *Elysia* sp. 6 was found at Jesser Point, Sodwana Bay National Park.

52. ***Elysia* sp. 7**

Taxonomy: This species of *Elysia* appears to be undescribed. It is characterized by its light green, papillate parapodia with yellow and black spots.

Natural History: This species was found in the rocky intertidal area on the green alga, *Caulerpa racemosa*, on which it feeds. The egg mass is yellow and consists of several whorls.

Occurrence and Distribution: The single specimen was found at Mbibi, 20 km north of Sodwana Bay National Park.

Family Stiligeridae

53. ***Placida dendritica*** (Alder and Hancock, 1843)
Synonym: *Hermaea capensis* Macnae, 1954

Taxonomy: Macnae (1954*b*) described *Hermaea capensis* as being distinct from *Placida dendritica* on the basis of its radular teeth. However, re-examination of material from South Africa demonstrated that they do not differ significantly and the species are considered synonymous.

Natural History: *Placida dendritica* is most commonly found together with *Elysia viridis* on several species of green algae, including *Codium* spp., *Bryopsis* sp., and *Halimeda cuneata*.

Occurrence and Distribution: *Placida dendritica* has been found in southern Africa from Elands Bay, on the Atlantic coast, to Port Alfred, on the Indian Ocean. *Placida dendritica* appears to be cosmopolitan, as it has been found from the North Atlantic, the Mediterranean, Japan, Australia, and California.

Family Caliphyllidae

54. ***Aplysiopsis sinusmensalis*** (Macnae, 1954)

Taxonomy: This species was described by Macnae as a *Hermaeina*, but subsequent work has shown that *Aplysiopsis* is an older name. This species is characterized by its somewhat flattened cerata.

Natural History: *Aplysiopsis sinusmensalis* has been found on the green alga *Enteromorpha* sp. in estuaries and tidal swimming pools.

Occurrence and Distribution: Macnae described *Aplysiopsis sinusmensalis* from six specimens collected at Milnerton, just north of Cape Town. Subsequently, a large population of specimens was discovered at Buffels Bay, on the Indian Ocean side of the Cape Point Nature Reserve, and a single specimen was found in a high rock pool at Bloubergstrand. It appears to be endemic to southern Africa.

55. ***Phyllobranchillus orientalis*** (Kelaart, 1858)

Taxonomy: This species is distinguishable by its uniform green-brown coloration and flattened, leaf-like cerata.

Natural History: Specimens were collected in less than 1 m of water on rock surfaces.

Occurrence and Distribution: Three specimens were found in the tidal swimming pool at Salt Rock, on the Natal North Coast. The species appears to be nocturnal. *Phyllobranchillus orientalis* has been previously recorded from Ceylon, India, Japan, New Caledonia and Hawaii (present study). It is widely distributed throughout the Indo-Pacific.

56. ***Mourgona*** **sp.**

Taxonomy: The genus *Mourgona* differs from *Phyllobranchillus* in the shape of the buccal mass and in the male reproductive system. The present species is undescribed. It can be distinguished from *Phyllobranchillus orientalis* by the presence of minute black spots and papillate cerata.

Natural History: *Mourgona* sp. was found on a shallow reef in about 1 m of water.

Occurrence and Distribution: The single specimen was found at Mbibi, 20 km north of Sodwana Bay National Park and represents the first record of the genus from the Indo-Pacific. It has also been collected from Aldabra Atoll and Guam.

Order Notaspidea

Family Umbraculidae

57. ***Umbraculum sinicum*** (Gmelin, 1783)

Taxonomy: *Umbraculum sinicum* can be distinguished from all other South African opisthobranchs by its flat, plate-like, external shell (Fig. 4d).

Natural History: Macnae (1962*a*) noted that the animals are often half-buried in sand or mud during the day. Recently, specimens have also been observed crawling actively in rock pools during the day.

Occurrence and Distribution: Macnae (1962*a*) recorded *Umbraculum sinicum* from Mozambique, but recently specimens have been found as far south as Umgazana, Transkei. *Umbraculum sinicum* appears to be circumtropical.

Family Pleurobranchidae

58. ***Pleurobranchus inhacae*** Macnae, 1962

Taxonomy: *Pleurobranchus inhacae* is distinguished from the following species by its fewer, larger tubercles on the notum. It also has a deeper notch on the anterior portion of the notum.

Natural History: A single specimen was found under a rock in a tidal swimming pool. Macnae (1962*a*) commonly found this species in sea grass beds.

Occurrence and Distribution: *Pleurobranchus inhacae* is known only from southern Africa, from Salt Rock, Natal to Mozambique, although it may be more widespread.

59. ***Pleurobranchus nigropunctatus*** (Bergh, 1907)

Taxonomy: *Pleurobranchus nigropunctatus* can be distinguished from other species occurring in the region by means of its elongate papillae. This species of *Pleurobranchus* appears to be similar in appearance to *P. areolatus*, known from the Western Atlantic and Panamic Pacific. However, the two species differ in aspects of their reproductive anatomy.

Natural History: Specimens of this species have been found under rocks on shallow reefs from 5-10 m depth.

Occurrence and Distribution: *Pleurobranchus nigropunctatus* was originally described from Cape Infanta. Specimens have recently been collected from False Bay and Algoa Bay. The species appears to be endemic to southern Africa.

60. ***Pleurobranchus peroni*** (Cuvier, 1804)

Taxonomy: *Pleurobranchus peroni* is characterizied by the many small tubercles that cover the mottled notum.

Natural History: *Pleurobranchus peroni* has been found in rock pools at low tide. This species appears to be largely nocturnal and nothing is known about its feeding.

Occurrence and Distribution: Two specimens have been found, one at Sodwana Bay National Park, and the other at Park Rynie. Previously, Macnae (1962*a*) noted that this species was common at Inhaca, Mozambique. *Pleurobranchus peroni* is widely distributed throughout the Indo-Pacific.

61. ***Pleurobranchus xhosa*** Macnae, 1962

Taxonomy: This species can be distinguished from the two preceding species by a lack of tubercles within the reticulate pattern on the notum.

Natural History: This species was found in 2 m of water in a tidal swimming pool. No natural historical data were indicated by Macnae (1962*a*), but the single specimen subsequently collected was actively crawling at night.

Occurrence and Distribution: This species is known only from South Africa, where Macnae collected it from Port Alfred to Port St. John's. The single specimen collected in the present study was found farther north, at Salt Rock, Natal. This species may have been confused with *Pleurobranchus peroni* elsewhere in the Indo-Pacific.

62. ***Berthella tupala*** Marcus, 1957

Taxonomy: The South African specimen is similar to that described from Brazil as *Berthella tupala*. The translucent white notum with opaque white markings approximates the description by Marcus (1957).

Natural History: One specimen was found under a large rock, in the vicinity of white colonial tunicates.

Occurrence and Distribution: The single specimen was found in the mid-intertidal zone at Umgazana, on the northern Transkei coast. *Berthella tupala* is known from the western Atlantic, from Brazil to the Caribbean. The author has also collected this species in Hawaii.

63. ***Berthella plumula*** (Montagu, 1803)

Taxonomy: *Berthella plumula* is distinct from *B. tupala* in that it lacks opaque white markings on the notum. Internally, there are differences in the structure of the jaws and radula.

Natural History: Specimens are locally common under rocks in the intertidal region in association with sponges and tunicates. The egg mass is white and cylindrical, consisting of one to one and a half whorls.

Occurrence and Distribution: This species has been found along the Atlantic coast of the Cape Peninsula at Kommetjie and McClear's Beach and from the Indian Ocean at Knysna. *Berthella plumula* has previously been recorded from Northern Europe and the Mediterranean.

64. ***Berthella* sp.**

Taxonomy: *Berthella* sp. appears to be undescribed, differing from *P. tupala* and *B. plumula* by its coloration, jaws, and radula. The translucent white notum, with scattered, opaque, white spots, distinguishes this species.

Natural History: *Berthella* sp. has been found crawling on huge boulders at depths of 25 m.

Occurrence and Distribution: Only two specimens of *Berthella* sp. have been found. Both were collected from the Atlantic coast of the Cape Peninsula, at Bakoven. This species appears to be endemic to the Cape Province.

65. ***Berthellina citrina*** (Rüppell and Leuckart, 1828)
Synonyms: see Edmunds and Thompson (1972).

Taxonomy: This species was recorded by Macnae (1962*a*) as *Berthellina granulata* and *B. punctata*. Subsequently, both have been considered synonyms of *B. citrina*. It is easily recognized by its smooth dorsum and yellow to orange color, with or without opaque white spots. *Berthellina engeli*, from the eastern Pacific, is a distinct species. Its shell is situated more posteriorly than that of *B. citrina*.

Natural History: Macnae (1962*a*) noted that this species appears to be a scavenger, even though entire shrimp were found in its stomach. This species is most common in rock pools, but has been found down to 2 m. It is one of the most common intertidal opisthobranchs along the South African coast. The egg mass is an orange collar of about one whorl which stands on its edge.

Occurrence and Distribution: Specimens have been found from Kommetjie, on the Atlantic coast of the Cape Peninsula, to Mozambique. *Berthellina citrina* is found throughout the Indo-Pacific to the Hawaiian Islands.

Family Pleurobranchaeidae

66. ***Pleurobranchaea tarda*** Verrill, 1880
Synonym: *Pleurobranchaea capensis* Vayssiere, 1901

Taxonomy: *Pleurobranchaea* is clearly distinguishable from other notaspideans described here in that the rhinophores are well separated, rather than being joined basally under the notum. Several other species of *Pleurobranchaea* are known from South African waters. The taxonomy of the genus was, until recently, poorly understood.

Natural History: *Pleurobranchaea tarda* is sporadically common from the intertidal water to at least 60 m. It is a voracious predator on other opisthobranchs and has been observed to feed on smaller individuals of its species. It lays a cylindrical egg mass consisting of several whorls.

Occurrence and Distribution: Specimens have been dredged from the coast of the Cape Province and have been found while scuba diving from the Atlantic coast of the Cape Peninsula to Knysna. Intertidal specimens, with egg masses, have been collected at Dale Brook in False Bay and from Knysna. *Pleurobranchaea tarda* was described from the Atlantic coast of North America and has also been found in West Africa.

67. ***Pleurobranchaea bubala*** Marcus and Gosliner, 1984

Taxonomy: This species was recently described. It differs from *Pleurobranchaea tarda* by its larger size and by having large, opaque, white pustules on the notum. There are internal morphological differences, as well.

Natural History: *Pleurobranchaea bubala* is found in the same habitat as *P. tarda*. It has been collected from shallow subtidal waters at 5-20 m depth. Like the preceding species, it is a voracious carnivore, preying on other opisthobranchs.

Occurrence and Distribution: *Pleurobranchaea bubala* appears to be endemic to the Cape Province of South Africa, where it is seasonally common in False Bay during the winter. Specimens have been collected from Buffels Bay to Jeffrey's Bay.

Order Nudibranchia
Suborder Doridacea
Family Dorididae

68. ***Doris verrucosa*** Linnaeus, 1758
Synonym: *Archidoris granosa* Bergh, 1907

Taxonomy: Bergh (1907) described this species from South Africa under the name of *Archidoris granosa*, although he also recorded *Doris verrucosa* from South Africa in the same paper. This species is the only South African dorid with a uniform yellow-brown body and large tubercles.

Natural History: *Doris verrucosa* is found in high intertidal pools, where it feeds upon the sponge, *Hymeniacedon perlevis*. Its egg mass is a tall, stiff collar of several complete whorls.

Occurrence and Distribution: *Doris verrucosa* has been found along most of the South African coast, from Luderitz, Namibia, on the Atlantic coast, to Port Elizabeth on the Indian Ocean. *Doris verrucosa* is widely distributed in the eastern and western Atlantic and extends into the warm temperate waters of the Indian Ocean.

69. ***Doris* sp. 1**

Taxonomy: This species is brownish yellow with darker brown spots. As the radular morphology differs from that of *Doris verrucosa*, this species appears to be undescribed.

Natural History: Specimens have been collected under rocks on shallow subtidal reefs in association with an unidentified species of sponge.

Occurrence and Distribution: The two specimens collected were found in Algoa Bay. At present, this species is known only from southern Africa.

70. ***Doris* sp. 2**

Taxonomy: This undescribed species differs from most known species of *Doris* in its external appearance, radula, and reproductive morphology.

Natural History: *Doris* sp. 2 is often found on tropical shallow water reefs, where it feeds on sponges.

Occurrence and Distribution: *Doris* sp. 2 is known only from South African waters, at Sodwana Bay National Park. Outside southern Africa this species is known from Tanzania (W. Rudman, personal communication).

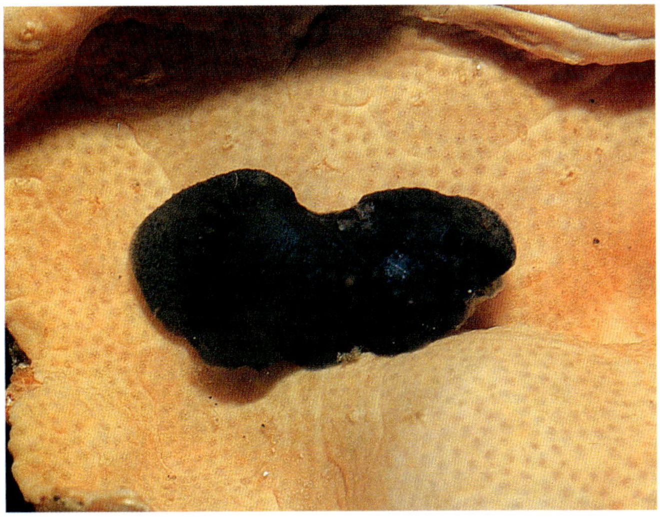

71. ***Doriopsis pecten*** (Collingwood, 1881)

Taxonomy: *Doriopsis pecten* (Fig. 11a), with gills arranged in a fan-like fashion rather than a circle, can be distinguished from all other South African dorids.

Natural History: *Doriopsis pecten* is found on a blue-black sponge on which it feeds. The coloration of the nudibranch and its egg mass are virtually identical to that of its prey species.

Occurrence and Distribution: Two specimens from the tidal swimming pool at Salt Rock, Natal, represent the first record of this species from the Indian Ocean. *Doriopsis pecten* has been recorded from Japan, Australia, New Caledonia, and Hawaii. It is widely distributed throughout the Indo-Pacific.

72. ***Atagema rugosa*** Pruvot-Fol, 1951

Taxonomy: This dorid has a distinctive mid-dorsal ridge, white coloration with black spots, and large tubercles.

Natural History: *Atagema rugosa* feeds on an unidentified demosponge.

Occurrence and Distribution: Only four specimens of this species have been found, all from 10 to 15 m depth at Oudekraal on the Atlantic coast of the Cape Peninsula. This species is previously known only from the Mediterranean.

73. ***Atagema gibba*** Pruvot-Fol, 1951

Taxonomy: This species is similar to *A. rugosa* but lacks the distinct mid-dorsal ridge and uniformly arranged tubercles. It also attains a larger body size.

Natural History: This species has been found under rocks on shallow subtidal reefs in about 10 m of water.

Occurrence and Distribution: The single specimen found in southern Africa was collected in Algoa Bay. Outside of southern Africa this species is known from England to the Mediterranean.

Family Discodorididae

74. ***Discordoris fragilis*** (Alder and Hancock, 1864)

Taxonomy: Light grey coloration, with darker gray spots arranged in distinct rows, characterizes this discodorid.

Natural History: *Discodoris fragilis* feeds on an unidentified grey sponge under rocks in the mid-intertidal zone.

Occurrence and Distribution: Specimens collected from Port Elizabeth, Gonubie, Park Rynie, and Sodwana Bay National Park constitute the first records of this species from southern Africa. *Discodoris fragilis* is widely distributed in the Indo-Pacific, where it has been collected from Tanzania, the Red Sea, India, the Philippines, Samoa, and Hawaii. Most recently, *D. fragilis* has been recorded from the Canary Islands.

75. ***Discodoris* sp. 1**

Taxonomy: This species appears to differ from all previously described members of the genus *Discodoris.* It differs from all other South African dorids by its minutely granular notum.

Natural History: This species feeds on an unidentified demosponge.

Occurrence and Distribution: Only two specimens of this species have been collected. It has been found subtidally, on both sides of the Cape Peninsula, from Oudekraal and Venus Pool, and appears to be endemic to South Africa.

76. ***Discodoris* sp. 2**

Taxonomy: This discodorid appears to represent an undescribed species distinguishable by its tuberculate notum with brown and purple markings.

Natural History: This species inhabits shallow tropical reefs, where it, presumably, feeds on sponges. A larval tape worm emerged from the anus of this species soon after its was collected.

Occurrence and Distribution: A single specimen was found in 15 m of water at Sodwana Bay National Park.

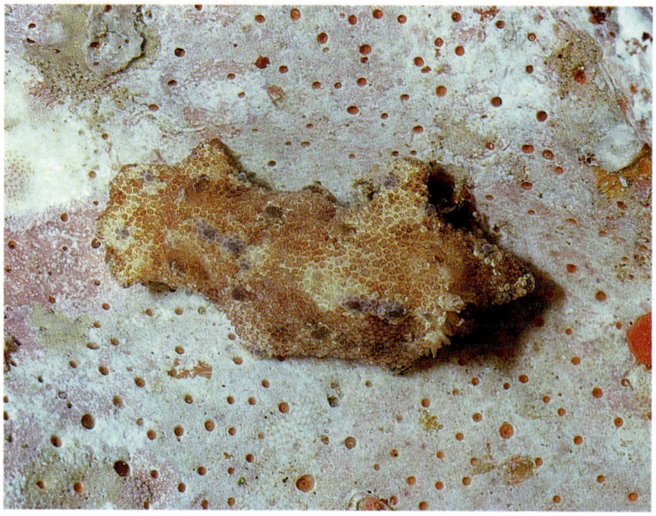

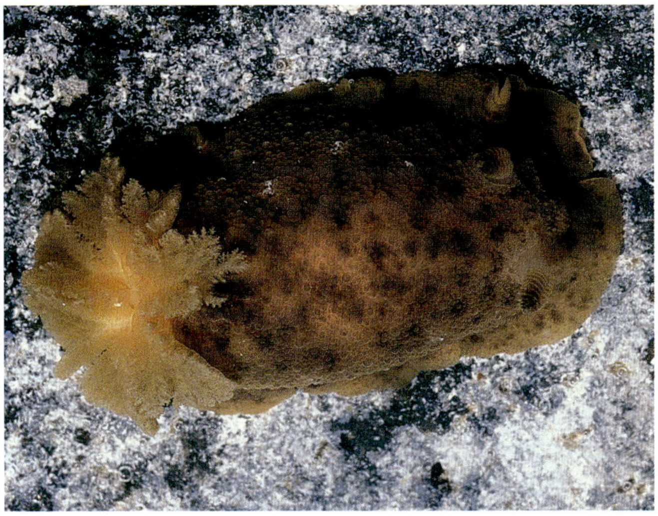

77. ***Geitodoris capensis*** Bergh, 1907

Taxonomy: *Geitodoris capensis* can be recognized by its orange color, brown spots, and simple, rounded tubercles. The gills are highly branched. The genus *Geitodoris* can be identified only by internal structures of the jaws, radula, and reproductive system.

Natural History: *Geitodoris capensis* has been collected most recently in intertidal rock pools but previously had been dredged. The intertidal specimen was associated with sponges.

Occurrence and Distribution: This species was described from a single specimen collected off Glendower Beacon, near Port Alfred. Recently, a single living specimen was collected near Kommetjie, on the Atlantic coast of the Cape Peninsula. This species appears to be endemic to southern Africa.

78. ***Anisodoris* sp. 1**

Taxonomy: This species appears to be undescribed. Its brown color with darker tubercles is distinctive. The genus is characterized by features of its internal anatomy, particularly the jaws and reproductive system.

Natural History: *Anisodoris* sp. 1 was found together with *Sclerodoris apiculata* in the mid-intertidal region.

Occurrence and Distribution: The single specimen was found at Coffee Bay, Transkei. This species has not been reported outside southern Africa.

79. ***Anisodoris* sp. 2**

Taxonomy: This species can be easily recognized by its light orange body color and darker orange circles surrounded by white.

Natural History: *Anisodoris* sp. has been found under rocks on shallow subtidal reefs in about 10 m depth.

Occurrence and Distribution: The single specimen was collected in Algoa Bay.

80. ***Thordisa* sp. 1**

Taxonomy: This undescribed species can be recognized by its greenish ground color and conical, fleshy papillae. *Thordisa* sp. 1 differs from the two *Thordisa* species described from southern Africa in its coloration and radular morphology.

Natural History: *Thordisa* sp. 1 has been found on dead coral reefs in less than 2 m depth and also from intertidal rock pools.

Occurrence and Distribution: Four specimens of this species have been collected in and around Sodwana Bay National Park. This species has not been collected outside southern Africa, although it is likely more widely distributed.

81. ***Thordisa* sp. 2**

Taxonomy: The lemon yellow ground color and larger body size distinguish this species from *Thordisa* sp. 1; *Thordisa* sp. 2 also appears to be undescribed.

Natural History: *Thordisa* sp. 2 was collected in an intertidal rock pool on a limestone reef flat.

Occurrence and Distribution: *Thordisa* sp. 2 was collected together with *T.* sp. 1 at Jesser Point, Sodwana Bay National Park and is known only from this single specimen.

Family Asteronitidae

82. ***Aphelodoris brunnea*** Bergh, 1907

Taxonomy: Members of the genus *Aphelodoris* possess a smooth notum and closely resemble the brightly colored chromodorids in body form. However, they are generally rather drably colored with white and brown pigment. *A. brunnea* is distinguishable by its white and brown pigment, arranged in a mottled pattern.

Natural History: *Aphelodoris brunnea* is found from the littoral zone to depths of about 15 m. It probably feeds on sponges. *A. brunnea* exhibits an escape response when disturbed and swims by flexing its body dorsoventrally.

Occurrence and Distribution: *Aphelodris brunnea* is endemic to South Africa and has been recorded from Buffels Bay, near Cape Point, to East London. It is sporadic in its occurrence except at Port Elizabeth, where it is common.

83. ***Aphelodoris* sp.**

Taxonomy: *Aphelodoris* sp. appears to be undescribed. It can be distinguished externally from *A. brunnea* by its uniform white coloration and few, large, dark brown spots.

Natural History: *Aphelodoris* sp. has been found in 20 to 30 m of water. Observation of the stomach contents of specimens shows that it feeds on sponges.

Occurrence and Distribution: Three specimens of *Aphelodoris* sp. have been collected at Llandudno, on the Atlantic coast and several specimens have been collected from Danger Point in the Indian Ocean. It appears to be endemic to southern Africa.

84. ***Sclerodoris apiculata*** (Alder and Hancock, 1864)

Taxonomy: The genus *Sclerodoris* has recently been reviewed by Rudman (1978). The boundaries of the genus are somewhat nebulous, but most species are characterizied by an irregularly tuberculate notum. The South African specimen of *Sclerodoris apiculata* closely approximates those previously described except that some tubercles have many fleshy papillae extending from them rather than a single papilla.

Natural History: *Sclerodoris apiculata* is found in the mid-intertidal region, where it feeds on sponges.

Occurrence and Distribution: A single specimen was found in southern African waters at Coffee Bay, Transkei. Elsewhere, *Sclerodoris apiculata* is known from Tanzania and southern India.

85. ***Sclerodoris coriacea*** Eliot, 1904

Taxonomy: *Sclerodoris coriacea* is similar in appearance to *S. apiculata* but has no fleshy extensions from its tubercles. The two species also differ in the anatomy of the radula and reproductive system.

Natural History: This species was found at night in a rock pool.

Occurrence and Distribution: The single specimen found in southern Africa was collected at Jesser Point, Sodwana Bay National Park. *Sclerodoris coriacea* was described from Zanzibar and may also be known from the East Indies, although the identification of these specimens is uncertain.

86. ***Sclerodoris* sp.**

Taxonomy: *Sclerodoris* sp. appears to be undescribed as its pattern of coloration is distinct from all described species.

Natural History: This species inhabits shallow, near-shore, tropical reefs.

Occurrence and Distribution: Three specimens of *Sclerodoris* sp. have been collected from Sodwana Bay National Park.

87. ***Artachaea* sp.**

Taxonomy: *Artachaea* sp. differs from other South African dorids by its regularly arranged tubercles (Fig. 10a) and brownish color. It appears to be undescribed. Only one species of *Artachaea* has been described. *Artachaea rubida* differs from *Artachaea* sp. in its coloration and radula.

Natural History: *Artachaea* sp. has been found intertidally at the sand-rock interface, where it feeds on sponges.

Occurrence and Distribution: Only two specimens of this species have been found, both from Knysna Lagoon. These specimens represent the first record of the genus from the Indian Ocean.

88. ***Halgerda wasinensis*** Eliot, 1904

Taxonomy: The genus *Halgerda* is closely allied to *Sclerodoris* but has distinctively elevated ridges which differ in color from the general body color. Also, the gills of *Halgerda* are not as highly branched. Rudman (1978) has shown that there are internal differences between *Halgerda* and *Sclerodoris*. *Halgerda wasinensis* is characterized by its dark brown to black notum with yellow to orange ridges.

Natural History: *Halgerda wasinensis* is one of the most common nudibranchs on subtidal tropical reefs along the northern Natal coast, where it feeds on sponges.

Occurrence and Distribution: *Halgerda wasinensis* has been found at Sodwana Bay National Park in 10 to 20 m of water. This represents the first record of *Halgerda wasinensis* from southern Africa. It has been previously recorded from Tanzania and Kenya.

89. ***Halgerda formosa*** Bergh, 1880

Taxonomy: *Halgerda formosa* is similar in appearance to *H. punctata* but differs in having dark black markings over the entire notum rather than just at the margin. This species has not been collected since 1880 and its relationship to *H. punctata* was subject to question.

Natural History: This species is found on shallow tropical reefs in about 5 m of water.

Occurrence and Distribution: The single specimen was collected at the entrance to Durban Harbor. *Halgerda formosa* is also known from single specimens collected at the type locality, Reunion Island, and from Mauritius.

90. ***Halgerda punctata*** Farran, 1902

Taxonomy: *Halgerda punctata* is characterized by its translucent white color with yellow-orange ridges and dark plum colored spots on the edges of the notum.

Natural History: This species inhabits shallow-water reefs dominated by dead coral. Little is known about its habits.

Occurrence and Distribution: The single southern African specimen was found at Adlam's Reef, Sodwana Bay National Park. Outside of southern Africa, *Halgerda punctata* is known from the type locality, Ceylon, and from a single specimen collected off Zanzibar.

Family Actinocyclidae

91. ***Hallaxa* sp.**

Taxonomy: *Hallaxa* sp. is recognizable by its pale yellow color and minute, opaque, white tubercles. Its color and radular morphology distinguish it from the four previously described species of *Hallaxa*.

Natural History: This species was found on a yellow sponge underneath a small rock in 2 m of water. It is very cryptic and is difficult to distinguish from its sponge prey.

Occurrence and Distribution: Two specimens were collected together at Vetchies Pier, just north of the entrance to Durban Harbor.

Family Kentrodorididae

92. ***Jorunna zania*** Marcus, 1976

Taxonomy: *Jorunna zania* was previously confused with *J. funebris* (Kelaart, 1859), but there are several external and internal differences between the two species. *J. zania* is distinguishable from all other South African dorids by its white ground color with large black spots on the dorsum.

Natural History: *Jorunna zania* is found from the lower intertidal to shallow subtidal zones to at least 10 m depth. It feeds on brownish sponges. Its egg mass is a high, undulating collar of one and a half to two whorls.

Occurrence and Distribution: *Jorunna zania* has been found in South African waters along the Natal coast from Durban Haror to Sodwana Bay. *Jorunna zania* was previously known only from Tanzania.

93. ***Jorunna tomentosa*** (Cuvier, 1804)

Taxonomy: Marcus (1976) has shown that several forms, previously thought to constitute one cosmopolitan species, are actually morphologically distinct. The South African specimens most closely resemble the material described from European waters as *J. tomentosa.*

Natural History: *Jorunna tomentosa* feeds on the sponge, *Haliclona* sp. from the mid-intertidal to about 10 m in depth.

Occurrence and Distribution: *Jorunna tomentosa* has been found along the Atlantic coast of South Africa,from Elands Bay to as far east as Knysna, on the Indian Ocean. The species is recorded only from Europe and southern Africa. The biogeography of the various species of *Jorunna*, once considered as *J. tomentosa*, requires more detailed study of large populations.

94. ***Gargamella* sp. 1**

Taxonomy: *Gargamella* sp. 1 is the only South African dorid with caryophyllidia on the notum that has a dark grey ground color. It most closely resembles *Jorunna tomentosa*, but can be distinguished by its dark grey spots, each surrounded by a white circle, and by its radular morphology.

Natural History: This species was found in the lower intertidal region under a rock covered by a grey sponge.

Occurrence and Distribution: The single specimen was collected from Dale Brook on the western shore of False Bay.

95. ***Gargamella* sp. 2**

Taxonomy: *Gargamella* sp. 2 also appears to be undescribed. It can be distinguished from *J. tomentosa* and *G.* sp. 1 by its white color. *Gargamella* sp. 2 can also be differentiated from *J. zania* by its brown, rather than black, spots, and by its uniformly white gills.

Natural History: This species was found under an overhang in 15 m of water.

Occurrence and Distribution: The single specimen was collected at Buffels Bay, on the False Bay coast of Cape Point.

Family Rostangidae

96. ***Rostanga muscula*** (Abraham, 1877)

Taxonomy: *Rostanga muscula* is characterized by a unique pattern of denticulation of radular teeth, as are other species of this genus. Externally, it is characterized by a red-orange color with darker mottling on the central portion of the notum.

Natural History: The majority of described species of *Rostanga* feed on red or orange sponges of the genera *Ophlitaspongia*, *Microciona*, and *Holoplocamia*. Both specimens found in South Africa were under rocks in the mid-intertidal zone, near red sponges.

Occurrence and Distribution: In the waters of southern Africa, *Rostanga muscula* is known only from Park Rynie, on the Natal South coast, and Sodwana Bay National Park. *Rostanga muscula* appears to be distributed throughout the Indo-Pacific tropics, although collections have been made only from South Africa, Australia, New Zealand, and Japan.

97. ***Rostanga* sp. 1**

Taxonomy: *Rostanga* sp. 1 (Figs. 9a, 10b) appears to be undescribed and differs from all described species by its yellow-orange to red coloration with black spots.

Natural History: *Rostanga* sp. 1 is found from the mid-intertidal region to a depth of at least 30 m. Intertidal specimens are associated with red sponges and are, therefore, bright red-orange in color. Subtidal specimens feed on different sponges and often possess yellow-orange pigment.

Occurrence and Distribution: This species has been found from the Atlantic coast of the Cape Peninsula at Oudekraal to Rooi Els, beyond the eastern shore of False Bay. It has not been recorded outside southern Africa.

98. ***Rostanga* sp. 2**

Taxonomy: Like the preceding species, *Rostanga* sp. 2 is undescribed. It is the only known species of the genus with a white body.

Natural History: *Rostanga* sp. 2 is found in shallow subtidal waters from 10 to 20 m. It feeds on a large, branching species of white sponge.

Occurrence and Distribution: *Rostanga* sp. 2 has been found only along the Atlantic coast of the Cape Peninsula from Oudekraal and Llandudno.

99. ***Rostanga* sp. 3**

Taxonomy: Like the preceding two species, *Rostanga* sp. 3 is also undescribed. It can be distinguished from all other species by its orange body with pale white patches.

Natural History: This species is found on the undersides of rocks on shallow subtidal reefs. It is associated with an unidentified species of orange sponge.

Occurrence and Distribution: Several specimens have been found in Algoa Bay. *Rostanga* sp. 3 has not been found outside of southern Africa.

Family Aldisidae

100. ***Aldisa trimaculata*** Gosliner, in Millen and Gosliner, 1985

Taxonomy: This recently described species (Fig. 9 c) differs from all other members of the genus by its tan coloration with three brown spots.

Natural History: Little is known about the natural history of this species except that it feeds on sponges in 15 to 30 m of water.

Occurrence and Distribution: This species has been found along the Atlantic coast of the Cape Peninsula, from Bakoven and Llandudno, and appears to be endemic to southern Africa.

101. ***Aldisa benguelae*** Gosliner, in Millen and Gosliner, 1985

Taxonomy: *Aldisa benguelae* differs from all other members of the genus by its uniformly orange body color, aspects of its radula, and reproductive system.

Natural History: *Aldisa benguelae* has been found in 20 to 30 m of water on red sponges, but little is known of its natural history.

Occurrence and Distribution: This species appears to be quite rare, as only three specimens have been found. It is known from the Atlantic coast of the Cape Peninsula, at Bakoven and Oudekraal, and is known only from southern Africa.

Family Platydorididae

102. ***Platydoris cruenta*** (Quoy and Gaimard, 1832)

Taxonomy: *Platydoris cruenta* is characterized by its white notum with many thin, undulating brown lines.

Natural History: This species was found under rocks in 2 m of water.

Occurrence and Distribution: The single specimen was found at the entrance to Durban Harbor. *Platydoris cruenta* is known from several localities throughout the Indo-Pacific, but has not been previously recorded from the western Indian Ocean.

103. ***Platydoris* sp.**

Taxonomy: This species appears to be undescribed and differs from the preceding species and *P. scabra*, which have also been recorded from southern Africa. It is white with orange rhinophores and gills and scattered dark brown blotches.

Natural History: The single specimen was trawled from deep water.

Occurrence and Distribution: This species is known only from the Transkei coast.

Family Chromodorididae

104. ***Chromodoris hamiltoni*** Rudman, 1977

Taxonomy: The systematics of the species complex to which *Chromodoris hamiltoni* belongs has been recently reviewed (Rudman 1977). *C. hamiltoni* is unique among these species, with its blue ground color and orange gills and rhinophores.

Natural History: *Chromodoris hamiltoni* is the most common nudibranch on subtidal tropical reefs in southern Africa. It has been found in shallow waters from 10 to 20 m deep. It feeds on sponges, but is commonly found crawling conspicuously over the living reef.

Occurrence and Distribution: *Chromodoris hamiltoni* has been found along the Natal coast, from Aliwal Shoals to Sodwana Bay National Park. This species was previously known from Tanzania, Kenya, and Madagascar.

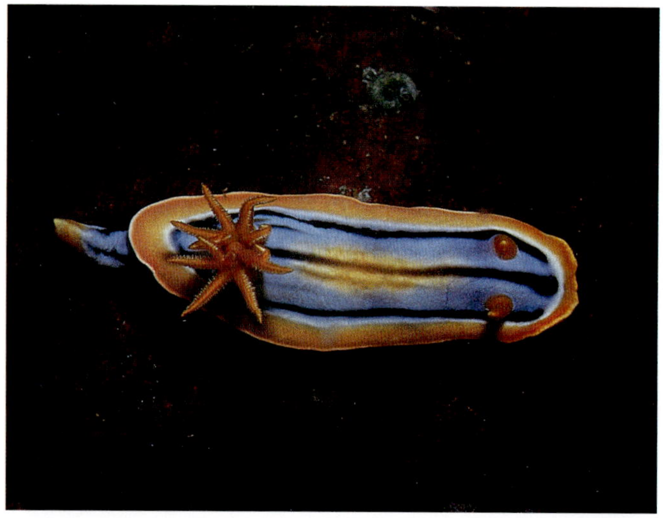

105. ***Chromodoris africana*** Eliot, 1904

Taxonomy: This species is very closely allied to the preceding species. Its black ground color, white lines, and orange rhinophores and gills readily distinguish it from other chromodorids.

Natural History: *Chromodoris africana* is found together with *C. hamiltoni* but is far less common. It also attains a larger body size.

Occurrence and Distribution: In southern Africa, *Chromodoris africana* is known only from subtidal reefs at Sodwana Bay National Park. This represents the first record of this species from southern Africa. It has been previously recorded from the Red Sea and the western Indian Ocean.

106. ***Chromodoris annulata*** Eliot, 1904

Taxonomy: *Chromodoris annulata* is identified by its white ground color, yellow-orange spots, and two large purple rings, one surrounding the gills and the other surrounding the rhinophores.

Natural History: Several specimens were found in rock pools at low tide.

Occurrence and Distribution: *Chromodoris annulata* has been found along the Natal coast, from Park Rynie to Sodwana Bay National Park. *Chromodoris annulata* is widespread in the Indo-Pacific. It has been found from the Red Sea, Kenya and Tanzania, South Africa, and recently from the Gulf of California.

107. ***Chromodoris inopinata*** Bergh, 1905

Taxonomy: *Chromodoris inopinata* is distinguishable from other South African dorids by its reticulate red markings and outer orange and inner maroon marginal lines. Rudman (1973) reviewed the *Chromodoris* species with this color pattern and concluded that *C. inopinata* is distinct.

Natural History: *Chromodoris inopinata* has been found on dead coral rubble at 2 m depth.

Occurrence and Distribution: The single specimen, found at Adlam's Reef, Sodwana Bay National Park, represents the first record of this species from southern Africa. Elsewhere, *Chromodoris inopinata* has been collected from Zanzibar and from Saleyar Island in Indonesia.

108. ***Chromodoris alderi*** Collingwood, 1881

Taxonomy: *Chromodoris alderi* has been much confused with other species with reticulate markings. It differs from either *C.* inopinata or *Chromodoris* sp. 6 by having the red reticulations restricted to the center of the notum.

Natural History: This species was found in a tidal swimming pool on the surface of a rock in 1 m of water.

Occurrence and Distribution: The single specimen was collected at Chaka's Rock, on the Natal North coast. It is known also from Formosa and Japan.

109. ***Chromodoris vicina*** Eliot, 1904

Taxonomy: This species of *Chromodoris* can be distinguished by a notum with yellow-orange mottling, black spots surrounded by an opaque white halo, and a marginal purple ring.

Natural History: *Chromodoris vicina* was found on a rocky tropical reef among sponges in about 1 m of water.

Occurrence and Distribution: Three specimens have been collected in South Africa at Sodwana Bay National Park. Prior to its discovery in South African waters, *Chromodoris vicina* was known only from Tanzania.

110. ***Chromodoris marginata*** (Pease, 1860)

Taxonomy: *Chromodoris marginata* is characterized by its white notum, rhinophores, and gills and yellow and orange marginal lines.

Natural History: This species is one of the more common intertidal nudibranchs on the tropical coast of South Africa. It inhabits shallow rock pools and surge channels.

Occurrence and Distribution: Specimens have been collected in southern Africa from Park Rynie and Sodwana Bay National Park. *Chromodoris marginata* is an Indo-Pacific species known from India, Vietnam, Japan, Indonesia, Australia, and Hawaii.

111. ***Chromodoris geometrica*** (Risbec, 1928)

Taxonomy: *Chromodoris geometrica* can be distinguished by its whitish-yellow pustules, a purple reticulate pattern, and reddish gills and rhinophores.

Natural History: This species was found on dead limestone reefs in 1 to 2 m depth.

Occurrence and Distribution: Four specimens, collected at Adlam's Reef, Sodwana Bay National Park, constitute the first record of *Chromodoris geometrica* from southern Africa. This species is widely distributed in the Indo-Pacific where it has been collected from Kenya, New Caledonia, Australia, and Enewetak Atoll.

112. ***Chromodoris* sp. 1**

Taxonomy: This species is undescribed. It is unique by virtue of its translucent white coloration, usually with large red spots, and an opaque white line on the edge of the notum. It may also have an inner yellowish-orange line. It is one of the most variably colored dorids in southern Africa.

Natural History: *Chromodoris* sp. 1 is commonly found on rocky reefs in the temperate waters of the Cape Province. It feeds on sponges and has been found from the intertidal fringe to depths of at least 30 m.

Occurrence and Distribution: *Chromodoris* sp. 1, *C.* sp. 2 and *Hypselodoris carnea*, are the only species of chromodorids that have been found in the cold temperate waters off the Atlantic coast of the Cape Peninsula. *Chromodoris* sp. 1 has been found subtidally from the Cape Peninsula to Port Elizabeth. Intertidal specimens have been found at Cape Hangklip. This species appears to be endemic to the Cape Province.

113. ***Chromodoris*** sp. 2

Taxonomy: This species appears to be undescribed and is characterized by uniform pink coloration with opaque white glands along the margin of the notum. Specimens reach a maximum of 15 mm in length.

Natural History: *Chromodoris* sp. 2 has been found on a pink encrusting sponge in 10 m of water.

Occurrence and Distribution: About 10 specimens of this species have been found from a single collection at Oudekraal, on the Atlantic coast of the Cape Peninsula.

114. ***Chromodoris* sp. 3**

Taxonomy: This species could not be identified as any described species. It is characterized by a yellow notum and rhinophores, a grey-purple submarginal ring, and an opaque white marginal ring. The body is decorated by three rows of black spots, each surrounded by an opaque white ring. The gills are opaque white with a rose colored central axis.

Natural History: The specimen was found under a dead coral head on a tropical reef in 15 m of water.

Occurrence and Distribution: The single specimen was found at Sodwana Bay National Park. This species has also been collected from Tanzania.

115. ***Chromodoris* sp. 4**

Taxonomy: This species closely resembles *Chromodoris decora* Pease, 1860, but differs in possessing opaque white spots on the notum. Also, the sub-marginal band has a purplish tinge, rather than the opaque white one found in *C. decora*. Until additional material from South Africa is available, the two species are regarded as distinct.

Natural History: The specimen was collected in a shallow rock pool at low tide.

Occurrence and Distribution: The single specimen was collected at Sodwana Bay National Park. This species has also been found in the Seychelles.

116. ***Chromodoris* sp. 5**

Taxonomy: This species appears to be distinct from all described species. It is a light rose pink with irregular, opaque white mottling on the notum.

Natural History: *Chromodoris* sp. 5 is found in the same habitat as the preceding species, on shallow water tropical reefs.

Occurrence and Distribution: A single specimen of this species was found at Sodwana Bay National Park. This species has also been found in the Seychelles.

117. ***Chromodoris* sp. 6**

Taxonomy: This species closely resembles *Chromodoris inopinata* but lacks an inner reddish purple line with opaque white spots. It also differs by having red spots on the hind end of the foot, and probably represents a distinct, undescribed species.

Natural History: *Chromodoris* sp. 6 has been found from intertidal rock pools to a depth of 15 m on living reefs.

Occurrence and Distribution: Two specimens have been collected from Umgazana, Transkei and from Sodwana Bay National Park.

118. ***Chromodoris* sp. 7**

Taxonomy: Although the present material is similar to *Chromodoris albopustulata* (Pease, 1860), it has much less orange pigment on the back and probably represents a distinct, undescribed species.

Natural History: Specimens have been found in the intertidal and shallow subtidal zones under rocks, dead coral heads, and on rock overhangs. It appears to be associated with sponges.

Occurrence and Distribution: *Chromodoris* sp. 7 has been found in southern Africa from Umgazana, Transkei to Sodwana Bay National Park. *Chromdoris* sp. 7 has also been found in Tanzania.

119. ***Chromodoris* sp. 8**

Taxonomy: *Chromodoris* sp. 8 appears to be undescribed and can be distinguished by orange rhinophores, gills, an interrupted notal margin, and its central dark purple spots.

Natural History: This species was found in a lower intertidal rock pool.

Occurrence and Distribution: The single specimen was collected at Park Rynie, Natal.

120. ***Durvilledoris lemniscata*** (Quoy and Gaimard, 1832)

Taxonomy: The present material and that recorded from Tanzania by Eliot (1904), differ from Bergh's description in that there is a central white band with paired, yellow lateral bands. In Bergh's specimen there was a central yellow band and lateral white bands. Nonetheless, they are considered here to represent the same species.

Natural History: *Durvilledoris lemniscata* has been found in intertidal rock pools.

Occurrence and Distribution: A single specimen was collected from Park Rynie on the Natal South coast. *Durvilledoris lemniscata* is known from widespread localities throughout the Indo-Pacific, namely Tanzania, Indonesia, and Tahiti.

121. ***Noumea varians*** (Pease, 1871)

Taxonomy: *Noumea* is closely allied to *Chromodoris*, but differs in the shape of its radular teeth. The present material agrees entirely with that described by Pease. Its rose-violet body, with two opaque white spots and yellow marginal line, clearly distinguishes this species.

Natural History: Specimens were found under stones or crawling in the open in intertidal rock pools.

Occurrence and Distribution: *Noumea varians* has been found in South Africa only at Sodwana Bay National Park. The occurrence of *Noumea varians* in South Africa represents the first record of this species from the Indian Ocean. It is known from the Society Islands and has recently been found at Guam.

122. ***Noumea purpurea*** (Baba, 1949)

Taxonomy: The body color and structure of the radula closely match that described by Baba. There is little doubt that the present material is conspecific with that from Japan.

Natural History: This species has been found in 15 m of water on tropical reefs, where it inhabits the under surface of coral rubble.

Occurrence and Distribution: In South African waters, *Noumea purpurea* is known from two specimens collected at Sodwana Bay National Park. Outside South Africa, *Noumea purpurea* is known only from the type locality in Japan.

123. ***Noumea* sp.**

Taxonomy: The general body color is white or pink. The gills, rhinophores, and an interrupted, marginal band are red-orange. This species is similar to *Noumea decussata* but differs in its radular morphology and is considered to be undescribed.

Natural History: Specimens were found in association with sponges under a rock in the mid-intertidal region.

Occurrence and Distribution: The two specimens that have been found were under the same rock at Umgazana, Transkei.

124. ***Thorunna horologica*** Rudman, 1984

Taxonomy: This species (Fig. 6c) is similar to *Chromodoris marginata* in that the ground color is white with a red-orange marginal ring. However, the gills and rhinophores are red-orange, rather than white, and an additional blue marginal line is present around the edge of the foot. Internally, the radular morphology differs significantly from other southern African chromodorids.

Natural History: *Thorunna horologica* is found under rocks on shallow tropical reefs in less than 10 m of water.

Occurrence and Distribution: Specimens have been found at Vetchies Pier near the northern entrance to Durban Harbor and at Sodwana Bay National Park. Outside southern Africa, this species is known from Tanzania, the type locality.

125. ***Hypselodoris capensis*** (Barnard, 1927)

Taxonomy: The genus *Hypselodoris* differs from *Chromodoris* in its radular morphology. Externally, the animals tend to have a higher profile to the body. *Hypselodoris capensis* is similar to *H. carnea* but differs in its radular structure. The two species differ slightly in their coloration, with *H. capensis* possessing much more pink or brownish pigment and having a discontinuous purple margin at the anterior end of the head.

Natural History: *Hypselodoris capensis* feeds on a light blue sponge and is found from the intertidal zone to at least 20 m.

Occurrence and Distribution: *Hypselodoris capensis* has been found from the tip of Cape Point to Umgazana, Transkei. It appears to be endemic to South Africa.

126. ***Hypselodoris carnea*** (Bergh, 1889)

Taxonomy: Bergh noted that in this species the majority of radular teeth are denticulate. *Hypselodoris carnea* was first recorded from South Africa by Eliot in 1910.

Natural History: *Hypselodoris carnea* has been found on dead coral and in rock pools to a depth of 10 m.

Occurrence and Distribution: This species has been found on the Natal coast from Durban Harbor to Sodwana Bay. *Hypselodoris carnea* is known only from South Africa and the type locality, Mauritius, although it is probably more widespread.

127. ***Hypselodoris infucata*** (Rüppell and Leuckart, 1828)

Synonyms: Rudman (1977). This species has previously been recorded from South Africa as *Glossodoris runcinata* (Bergh, 1877) and *Chromodoridella mirabilis* Eliot, 1905.

Taxonomy: This species appears to be variable in its coloration, which accounts for the many names that have been applied to it. *Chromodoridella mirabilis* Eliot, 1905 is likely to be an abnormal specimen of this species, in which the gills opened under the notum.

Natural History: *Hypseldoris infucata* is found in the intertidal and shallow subtidal habitats on tropical reefs.

Occurrence and Distribution: This species is known from the Natal coast, from Scottburgh to Sodwana Bay. It appears to be widely distributed throughout the Indo-Pacific.

128. ***Hypselodoris maridadilus*** Rudman, 1977

Taxonomy: Rudman (1977) studied this complex of closely allied *Hypselodoris* species and noted that *H. maridadilus* had been misidentified from South Africa as *Glossodoris lineata.*

Natural History: *Hypselodoris maridadilus* has been found in 2 m of water on limestone rock surfaces in the vicinity of a variety of sponges.

Occurrence and Distribution: A single specimen of the species was found at Sodwana Bay National Park. Previously, it had been recorded from Scottburgh, on the Natal South coast. This species is widespread in the Indo-Pacific and has been recorded from Tanzania, Japan, Hawaii, and Enewetak Atoll.

129. ***Hypselodoris* sp. 1**

Taxonomy: This species appears to be undescribed and differs from all recorded members of the genus. The alternating red and white lines on the notum are characteristic of this species.

Natural History: *Hypselodoris* sp. 1 was found under coral rubble in 2 or 3 m of water.

Occurrence and Distribution: The single specimen collected was found near the entrance to Durban Harbor.

130. ***Hypselodoris* sp. 2**

Taxonomy: This nudibranch could not be identified with any described species. It is distinguishable by its translucent white body with a cream sub-marginal band, cream mid-dorsal mottling, and minute black spots. There is an interrupted, purple, marginal band and the gills and rhinophores possess red pigment.

Natural History: A single specimen was found crawling in a rock pool in the mid-intertidal region.

Occurrence and Distribution: The single specimen was collected at Park Rynie on the Natal South coast.

131. ***Hypselodoris* sp. 3**

Taxonomy: This species appears to be undescribed and can be differentiated by the patch of opaque white in the center of its notum.

Natural History: This species has been collected under rocks in the lower intertidal fringe and from shallow subtidal reefs.

Occurrence and Distribution: Two specimens have been collected at Umgazana, Transkei and Port Elizabeth.

132. ***Risbecia pulchella*** (Rüppell and Leuckart, 1828)

Taxonomy: The present material resembles that described by Rüppell and Leuckart from the Red Sea. This species, along with several other species from other areas in the Indo-Pacific, form a species complex which is in need of considerable revision.

Natural History: This species was found under coral rubble in 3 m of water on arborescent sponges.

Occurrence and Distribution: A single specimen was found adjacent to the mouth of Durban Harbor and an additional three specimens were found at Sodwana Bay National Park. This species appears to be widely distributed throughout the tropical Indo-Pacific.

133. ***Glossodoris atromarginata*** (Cuvier, 1804)

Taxonomy: *Glossodoris* is separated from *Chromodoris* largely by its more or less permanently convoluted notal margin. *Glossodoris atromarginata* is distinguishable by its yellow to brown color and dark purple to black marginal ring.

Natural History: This species was found under a dead coral head in 15 m of water.

Occurrence and Distribution: The single specimen collected at Sodwana Bay National Park represents the first record of this species from South Africa. *Glossodoris atromarginata* is a widespread Indo-Pacific species and has been found in Tanzania, the Red Sea, India, Sri Lanka, Singapore, Philippines, Thailand, Vietnam, China, Japan, the Solomon Islands, Australia, and Tahiti.

134. ***Glossodoris* sp. 1**

Taxonomy: This species of *Glossodoris* appears to be undescribed and is distinguishable by its white ground color with opaque white pustules on the notum and a yellow marginal ring.

Natural History: *Glossodoris* sp. 1 was found in 3 m of water crawling on rocky reefs.

Occurrence and Distribution: A single specimen of this species was found adjacent to the mouth of Durban Harbor.

135. ***Glossodoris* sp. 2**

Taxonomy: *Glossodoris* sp. 2 is undescribed and is differentiated by its pinkish ground color, yellow submarginal, and black marginal rings.

Natural History: *Glossodoris* sp. 2 has been found crawling in the open on living coral reefs in 10 to 15 m of water.

Occurrence and Distribution: Only two specimens of this species have been collected, both from Sodwana Bay National Park.

136. ***Glossodoris* sp. 3**

Taxonomy: This species of *Glossodoris* can be distinguished by its blue-grey marginal ring and bluish brown notum with opaque white spots. It is similar in coloration to *G. cincta* (Bergh, 1889), but differs significantly in its radular morphology.

Natural History: *Glossodoris* sp. 3 inhabits coral rubble in 10 to 15 m of water on tropical reefs.

Occurrence and Distribution: A single specimen was found at Sodwana Bay National Park and represents the first record of this species from southern Africa.

137. ***Glossodoris* sp. 4**

Taxonomy: This species is the only known member of the genus that is uniformly white with an opaque white mantle margin.

Natural History: This species was found crawling on the upper surface of a rock in about 10 m of water on a subtidal rocky reef.

Occurrence and Distribution: The single specimen of *Glossodoris* sp. 4 was collected in Algoa Bay. It is not known outside southern Africa.

138. ***Ceratosoma* sp.**

Taxonomy: This species is undescribed and differs from all known members of the genus by its unique color pattern of bluish spots.

Natural History: Several specimens of *Ceratosoma* sp. have been collected from deep subtidal reefs in 30 to 40 m of water.

Occurrence and Distribution: This species has only been collected in the vicinity of Danger Point, near Gans Baai.

139. ***Cadlina* sp. 1**

Taxonomy: *Cadlina* (Fig. 10c) is closely allied to *Chromodoris* but differs by its large rachidian tooth and white ground color with yellow, black, or brown ornamentation. *Cadlina* sp. 1 appears to be undescribed.

Natural History: *Cadlina* sp. 1 was found in the lower intertidal or shallow subtidal regions to a depth of approximately 10 m. It is associated with, and feeds upon, sponges.

Occurrence and Distribution: This species has been found in False Bay from Buffels Bay, Castle Rocks, and Dalebrook, eastward to Rooi Els.

140. ***Cadlina* sp. 2**

Taxonomy: *Cadlina* sp. 2 appears to be undescribed and can be distinguished from the preceding species by its lack of brown pigment across the mid-dorsal region.

Natural History: *Cadlina* sp. 2 has been found in 15 to 30 m of water, where it is associated with sponges, on which it presumably feeds.

Occurrence and Distribution: Several specimens have been collected along the Atlantic coast of the Cape Peninsula at Bakoven, Oudekraal, and Llandudno.

141. ***Cadlina* sp. 3**

Taxonomy: This species can be differentiated by its grey-blue color with black spots and lighter blue circles. It is also undescribed.

Natural History: Specimens have been collected under rocks on shallow subtidal reefs, where they appear to be associated with encrusting sponges.

Occurrence and Distribution: The two specimens collected were both found in Algoa Bay. *Cadlina* sp. 3 appears to be endemic to southern Africa.

142. ***Cadlina* sp. 4**

Taxonomy: *Cadlina* sp. 4 can be readily identified by its brownish color with a darker brown "W" shaped swath on the notum. Like the preceding species, it is also undescribed.

Natural History: Specimens of this species have been found under rocks on shallow subtidal reefs.

Occurrence and Distribution: Several specimens of this species have been collected from Algoa Bay. *Cadlina* sp. 4 is endemic to southern Africa.

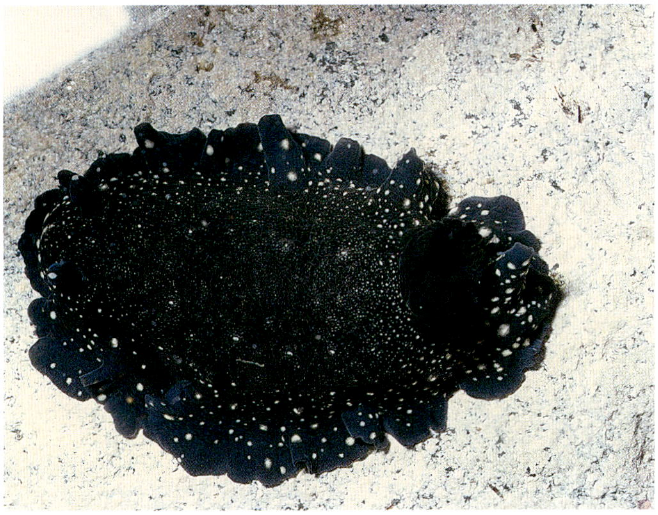

Family Hexabranchidae

143. ***Hexabranchus sanguineus*** (Rüppell and Leuckart, 1828)

Synonym: *Hexabranchus adamsi* Gray, 1850

Taxonomy: Thompson (1972) suggested that there is, perhaps, only one species in this genus. While some workers have questioned the synonymy of all members of the genus, Thompson is probably correct in combining the majority of species under *Hexabranchus sanguineus*. This species is immediately recognizable by its orange to red color and its large size, which can exceed 400 mm in length.

Natural History: One of the most remarkable things about *Hexabranchus sanguineus* is its ability to swim by undulating the margins of the notum. This species is fairly common on living coral reefs, from 10 to 20 m in depth. Although no observations of feeding were made, previous studies indicate that it feeds on sponges, tunicates, foraminiferans, worms, gastropods, and echinoderms.

Occurrence and Distribution: Specimens have been found in southern Africa, from Scottburgh (as *Hexabranchus adamsi*) to Sodwana Bay. This species is widespread throughout the Indo-Pacific and may, in fact, be circumtropical in its distribution.

Family Dendrodorididae

144. ***Dendrodoris nigra*** (Stimpson, 1855)

Taxonomy: *Dendrodoris nigra*, like other members of the genus, has a soft, slimy body. The color ranges from orange in juveniles to black in adults. A red ring around the margin of the notum and white spots may be present.

Natural History: Like most dendrodorids, this species feeds on sponges and is most commonly found in the intertidal zone. The egg mass is an orange, convoluted ribbon of several whorls.

Occurrence and Distribution: This species was recorded from Mozambique by Barnard (1927) but, more recently has been found as far south as Gonubie. *Dendrodoris nigra* is one of the most common intertidal nudibranchs in all tropical oceans.

145. ***Dendrodoris caesia*** (Bergh, 1907)

Synonym: *Dendrodoris kalkensis* (Barnard, 1927)

Taxonomy: The bluish and pink body, with a convoluted margin of the notum, characterizes *Dendrodoris caesia*. Bergh's description of this species is superficial. The fact that it was bluish when alive strongly suggests that Bergh's specimen is the same species as Barnard's *Doris kalkensis*. Barnard (1932) stated that the two species differ in the number of gills, but gill number varies considerably in *Dendrodoris*. As both appear similar and both are described from False Bay, they are considered to be synonymous.

Natural History: *Dendrodoris caesia* is found from the intertidal boundary to a depth of at least 20 m. It feeds on a yellow-brown sponge. Its egg mass is a convoluted ribbon of several whorls.

Occurrence and Distribution: *Dendrodoris caesia* has been found from Cape Point to as far east as Port Elizabeth. It can be exceedingly common, particularly along the south coast, where it may be the most common nudibranch. In Knysna Lagoon, in April 1980, hundreds of individuals were observed mating and laying eggs. *D. caesia* appears to be endemic to the Indian Ocean of the Cape Province.

146. ***Dendrodoris denisoni*** (Angas, 1864)

Taxonomy: Most species of *Dendrodoris* possess a smooth notum. *D. denisoni* is one of four species of *Dendrodoris* with tubercles on its back. It, nevertheless, is very soft to the touch. The brown notum with bright blue spots is another distinctive feature.

Natural History: *Dendrodoris denisoni* inhabits shallow tropical reefs.

Occurrence and Distribution: Specimens from Sodwana Bay National Park represent the first record of this species in southern Africa. More recently, it has been found as far south as Park Rynie. This species of *Dendrodoris* is widespread throughout the Indo-Pacific, extending as far east as Hawaii (Gosliner, unpublished data).

147. ***Dendrodoris* sp. 1**

Taxonomy: *Dendrodoris* sp. 1 appears to be undescribed. It is similar in general appearance to *D. nigra*, but differs by its greenish ground color and gills with black and opaque white mottling.

Natural History: This species was found underneath a stone in 2 m of water.

Occurrence and Distribution: The single specimen was collected at Vetchies Pier, just north of the entrance to Durban Harbor.

148. ***Dendrodoris* sp. 2**

Taxonomy: *Dendrodoris* sp. 2 appears to be undescribed. It is characterized by few pink tubercles, red gills and rhinophores, and black markings on the notum.

Natural History: This species was found crawling on a dead limestone reef in 2 m of water.

Occurrence and Distribution: The single specimen was collected at Mbibi, 20 km north of Sodwana Bay National Park.

149. ***Doriopsilla miniata*** (Alder and Hancock,1864)
Synonyms: *Doriopsilla areolata* Bergh, 1880
Doriopsilla clavisa Allan, 1933
Doriopsilla fedalae Pruvot-Fol, 1953

Taxonomy: Members of this genus are distinct from other dorids in that they lack a radula and a well developed head. *Doriopsilla* is rigid, with spicules embedded in the notum, while *Dendrodoris* is soft and fleshy. *Doriopsilla miniata* is distinguished by its yellow to orange body color with opaque white lines. *D. albolineata* Edmunds, 1968, and *D. janaina* Marcus and Marcus, 1967, are similar in appearance, but both appear to be distinct from *D. miniata*.

Natural History: *Doriopsilla miniata* feeds on sponges from the intertidal zone to a depth of at least 30 m. Though it lacks a radula, well developed oral glands permit external digestion of sponge cells, which are then pumped into the mouth.

Occurrence and Distribution: This species is widely distributed in South African waters, and has been found from the Atlantic coast of the Cape Peninsula to Sodwana Bay. *Doriopsilla miniata* is widely distributed, having been recorded from the Mediterranean Sea, India, Australia, and Japan.

150. ***Doriopsilla* sp. 1**

Taxonomy: *Doriopsilla* sp. 1 is white to brown with black spots on the notum. The rhinophores are orange and oblong, rather than round. This species is undescribed.

Natural History: *Doriopsilla* sp. 1 has been found in 10 to 400 m of water where it feeds on erect sponge colonies.

Occurrence and Distribution: This species is common along the Atlantic coast of the Cape Peninsula in subtidal communities. It appears to increase in abundance with increased depth. Recently specimens have been collected from Algoa Bay and Danger Point. *Doriopsilla* sp. 1 appears to be endemic to southern Africa.

151. ***Doriopsilla* sp. 2**

Taxonomy: The translucent white to tan body, often with a bluish tinge, opaque white tubercles, and a marginal notal ring, distinguishes this species of *Doriopsilla (Fig. 9b).*

Natural History: *Doriopsilla* sp. 2 has been found commonly among sponges in 10 to 30 m of water.

Occurrence and Distribution: This species has been found only along the Atlantic coast of the Cape Peninsula and appears to be restricted to the Atlantic waters of South Africa.

Family Phyllidiidae

152. ***Phyllidia varicosa*** Lamarck, 1801

Taxonomy: Although almost 50 species of *Phyllidia* have been described, *P. varicosa* is instantly recognizable by its blue color, yellow tubercles and rhinophores, and three longitudinal black lines.

Natural History: *Phyllidia varicosa* is found on subtidal coral reefs in 10 to 20 m. Little is known about feeding in members of the genus *Phyllidia*. Species of this genus produce toxic chemicals which are fatal to fish. They exude a characteristic pungent odor.

Occurrence and Distribution: *Phyllidia varicosa* has been found at Sodwana Bay National Park and Aliwal Shoals. It is widespread throughout the Indo-Pacific tropics.

153. ***Phyllidia* sp. 1**

Taxonomy: Specimens of this species are morphologically similar to *Phyllidia nobilis*. However, the arrangement of lines of tubercles, with a mid-doral, transverse interruption, distinguishes this species. As this characteristic is uniform in the several specimens found in South African waters, the material is considered distinct and probably represents an undescribed species.

Natural History: *Phyllidia* sp. 1 is found together with *P. varicosa* on tropical coral reefs but is somewhat less common. It has been found from 10 to 20 meters depth.

Occurrence and Distribution: This species has been found only at Sodwana Bay National Park.

154. ***Phyllidia* sp. 2**

Taxonomy: The pattern of longitudinal rows of compound tubercles, with a central, discontinous row, is unique to this apparently undescribed species.

Natural History: The species is found with the preceding two species on tropical coral reefs, but is, by far, the least common.

Occurrence and Distribution: *Phyllidia* sp. 2 has been collected only at Sodwana Bay National Park.

155. ***Ceratophyllidia africana*** Eliot, 1903

Taxonomy: The status of the genus *Ceratophyllidia* has been in question. Several workers have suggested that it should be regarded as a synonym of *Phyllidiopsis*. However, the presence of stalked, globular papillae, which can be shed, appears to sufficiently differentiate *Ceratophyllidia* as a distinct genus.

Natural History: *Ceratophyllidia africana* was found under a dead coral head in 18 m of water on a tropical reef.

Occurrence and Distribution: The South African specimen was found at Sodwana Bay National Park. Prior to its discovery in South Africa, this species was known only from the single specimen described by Eliot from Tanzania.

Family Onchidorididae

156. ***Acanthodoris* sp.**

Taxonomy: This undescribed species represents the first member of the Onchidorididae recorded from southern Africa. It is characterized by its brown notum with elongate papillae.

Natural History: *Acanthodoris* sp. has been found on the surface of shallow rocky reefs in association with bryozoans.

Occurrence and Distribution: Only two specimens of this species have been collected, both from Buffels Bay in False Bay. This species appears to be endemic to southern Africa.

157. ***Diaphorodoris* sp.**

Taxonomy: In its coloration, this undescribed species of *Diaphorodoris* resembles *Chromodoris marginata*, *Noumea* sp., and *Thorunna horologica*, but differs in having small tubercles on the notum and a mid-dorsal ridge on the foot. *Diaphorodoris* sp. also differs significantly in its internal morphology. *Diaphorodoris* sp. is similar to *D. mitsuii* Baba, 1938, but is white, rather than yellow, and possesses a ridge on the foot, which appears to be absent in *D. mitsuii*.

Natural History: Specimens have been found crawling in the open on dead reefs in depths of 1 to 2 m.

Occurrence and Distribution: The four specimens were collected at Adlam's Reef, Sodwana Bay National Park.

Family Goniodorididae

158. ***Okenia amoenula*** (Bergh, 1907)

Taxonomy: This species was first described by Bergh from the east side of False Bay. Subsequently, Macnae (1958) described much of the internal anatomy of the species. Elongate notal appendages distinguish this species.

Natural History: *Okenia amoenula* has been found from the intertidal zone to 30 m in rocky habitats.

Occurrence and Distribution: This species is generally uncommon but occasionally individuals have been encountered from the Atlantic coast of the Cape Peninsula, at Oudekraal to beyond Port Alfred. This species appears to be endemic to the Cape Province.

159. ***Okenia* sp.**

Taxonomy: This species is undescribed. It differs from *Okenia amoenula* in its coloration and by the presence of an elongate, mid-dorsal papilla just in front of the gills. It most closely resembles *O. impexa* Marcus,1957, but differs in its coloration and internal morphology.

Natural History: *Okenia* sp. was collected in 2 m of water under a small rock.

Occurrence and Distribution: The single specimen was found at Vetchies Pier, north of Durban Harbor.

160. ***Goniodoris mercurialis*** Macnae, 1958

Taxonomy: *Goniodoris mercurialis* is light brown or tan with opaque white spots. Macnae noted that it is distinct from *G. ovata* Barnard, 1934, on the basis of its radula. *G. mercurialis* lacks a mid-dorsal ridge between the rhinophores and gills; the ridge is present in *G. ovata*.

Natural History: This species feeds on compound tunicates and has been found under rocks in the intertidal region. The egg mass is a flattened spiral consisting of one whorl.

Occurrence and Distribution: This species was described from False Bay, but recently specimens have been found along the Atlantic coast of the Cape Peninsula at Bloubergstrand and Kommetjie. *Goniodoris mercurialis* appears to be endemic to the region surrounding the Cape Peninsula.

161. ***Goniodoris castanea*** Alder and Hancock, 1845
Synonym: *Goniodoris brunnea* Macnae, 1958

Taxonomy: Macnae described *Goniodoris brunnea* and noted that it was similar to *G. castanea*, although he indicated that there were radular differences. The similiarities in body form, however, clearly indicate that the two species are synonymous. The described radular differences are minor and within the range of variation of a single species.

Natural History: Like the preceding species, *Goniodoris castanea* is associated with, and feeds upon, compound tunicates. It has been found from the intertidal boundary to 15 m in depth. The egg mass is a spiral ribbon attached to the substrate by its thin edge and consists of two to four whorls.

Occurrence and Distribution: This species was described from South African waters from Kleinmonde, east of Port Alfred. Recently, specimens have been found on both coasts of the Cape Peninsula at Oudekraal, Buffels Bay, and Strandfontein. *Goniodoris castanea* is widespread, having been recorded from Europe, the Mediterranean, Japan, and New Zealand.

162. ***Goniodoris sp.***

Taxonomy: Only one 4-mm specimen of this species has been collected. This minute species appears to be undescribed, but inadequate material prevents a meaningful comparison with known taxa. It is clearly distinct from other South African species, with its white color and body shape.

Natural History: *Goniodoris* sp. inhabits intertidal rock pools and, presumably, feeds on compound tunicates.

Occurrence and Distribution: The single specimen was collected at Sodwana Bay National Park.

163. ***Trapania* sp. 1**

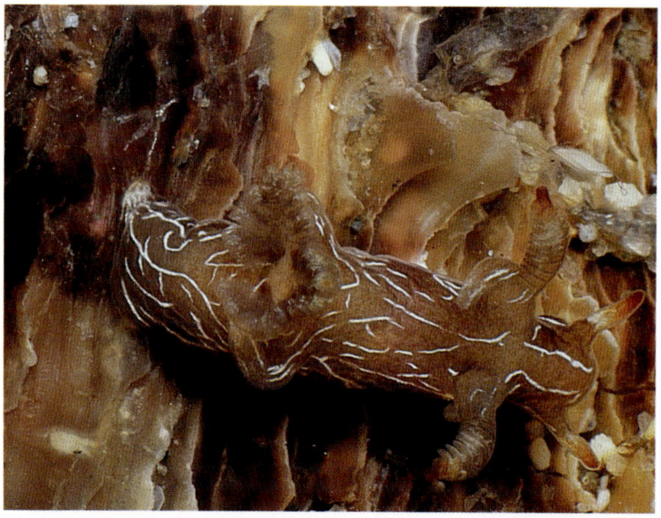

Taxonomy: The genus *Trapania* is characterized by curved extra-rhinophoral and extra-branchial processes. *Trapania* sp. 1 is similar to *Trapania lineata*, described from the Mediterranean, but differs in its color pattern and radular morphology. It is, consequently, considered to be a distinct, undescribed species.

Natural History: *Trapania* sp. 1 and its coiled ribbon of eggs have been found in the intertidal zone, where they are associated with ctenostomatous bryozoans.

Occurrence and Distribution: This species appears to be endemic to South Africa, having been found only from the Knysna Heads.

164. ***Trapania* sp. 2**

Taxonomy: *Trapania* sp. 2 is undescribed and can be distinguished from all described species by its greenish ground color with opaque white spots and yellow extra-rhinophoral and extra-branchial processes.

Natural History: This species was found floating upside down at the surface of an intertidal rock pool at night.

Occurrence and Distribution: The single specimen of *Trapania* sp. 2 was collected at Jesser Point, Sodwana Bay National Park.

165. ***Ancula* sp.**

Taxonomy: This undescribed species has a distinctive color pattern. *Ancula lentiginosa* Farmer and Sloan, 1964, is similarly colored, but has small brown spots rather than regular patches of brown.

Natural History: *Ancula* sp. has been found on arborescent bryozoans in 1 to 10 m of water.

Occurrence and Distribution: This species has been found along the Atlantic coast at Bloubergstrand and in False Bay at Castle Rocks. It appears to be endemic to South Africa.

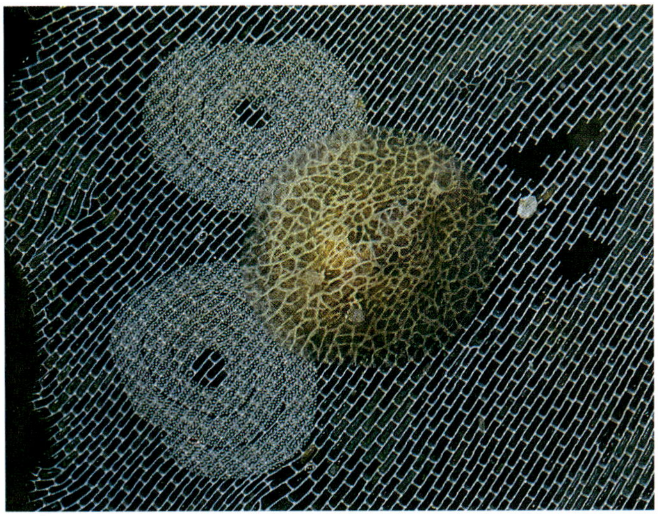

166. ***Corambe* sp.**

Taxonomy: The genus *Corambe* is characterized by a series of simple gills, situated in a posterior notch between the notum and foot. Since the species which occurs in South Africa has a distinct arrangement of gill leaflets, it is likely undescribed.

Natural History: *Corambe* sp. (Fig. 11b) is one of the most cryptic nudibranchs found in South Africa. It is associated with the bryozoan *Membranipora membranacea*, which forms large encrustations on broad-bladed algae. Its egg mass, also visible in the photograph, is similarly cryptic in appearance.

Occurrence and Distribution: *Corambe* sp. is sporadic in its occurrence. It has been found subtidally on kelp in about 10 m of water off Windmill Beach and Castle Rocks in False Bay. This species appears to be endemic to South Africa.

Family Polyceridae

167. ***Crimora* sp.**

Taxonomy: This species is similar in appearance to *Crimora lutea*, which is known from Japan and Australia. It differs slightly in its coloration and more significantly in its radular morphology, and is considered here to be undescribed.

Natural History: Specimens of *Crimora* sp. have been collected in 10 m of water on rocky subtidal reefs.

Occurrence and Distribution: Three specimens have been collected from Algoa Bay. *Crimora* sp. appears to be endemic to southern Africa.

168. ***Thecacera pennigera*** (Montagu, 1804)
Synonyms: *Thecacera maculata* Eliot, 1905
Thecacera lamellata Barnard, 1933

Taxonomy: *Thecacera pennigera* is white with orange and black spots. Willan (1976) combined *T. maculata* and *T. lamellata* with *T. pennigera*, since there are no significant internal differences and all three possess the same pattern of coloration. Several taxonomists have suggested that the widespread distribution of *T. pennigera* is a result of its being transported on ships' hulls. While this may account for many of the records from harbors, it does not explain the distribution of this species along the open coast of southern Africa, where specimens have been found several hundred kilometers from the nearest harbor.

Natural History: *Thecacera pennigera* feeds on arborescent bryozoans, particularly *Bugula* spp. Specimens have been found from the intertidal boundary to 15 m.

Occurrence and Distribution: *Thecacera pennigera* has been found from the Atlantic and Indian Ocean coasts of the Cape Peninsula, extending into Transkei and Natal. This species is widespread with a seemingly cosmopolitan distribution. It is known from Europe, Brazil, South Africa, Pakistan, Japan, Australia, and New Zealand.

169. ***Thecacera pacifica*** (Bergh, 1883)
Synonym: *Thecacera inhacae* (Macnae, 1958)

Taxonomy: *Thecacera pacifica* is distinguishable by its yellow to orange body color with black and blue markings. Recently, Burn (1974) discussed the taxonomy of this species and concluded that Bergh's genus *Ohola* is a synonym of *Thecacera*.

Natural History: This species has been found intertidally among sea grass, in 1 m of water, under rocks. Its food remains unknown.

Occurrence and Distribution: This species was previously known from southern African only from Macnae's collections at Inhaca, Mozambique. Recently, a single specimen was found near the entrance to Knysna Lagoon. *Thecacera pacifica* is widespread in the Indo-Pacific tropics, having been collected from Australia, the Arafura and the Red seas, as well as from southern Africa.

170. ***Thecacera* sp.**

Taxonomy: *Thecacera* sp. (Fig. 8a) possesses a unique color pattern and represents an undescribed species.

Natural History: Members of the genus *Thecacera* feed on arborescent bryozoans and *Thecacera* sp. is no exception. It has been found on bryozoan colonies in overhangs and ledges from the intertidal zone to 2 m depth.

Occurrence and Distribution: *Thecacera* sp. has been found from Umgazana, Transkei to Sodwana Bay National Park. It has been found only from South Africa, but may be more widespread.

171. ***Polycera capensis*** Quoy and Gaimard, 1824
Synonyms: *Polycera nigrocrocea* Barnard, 1927
Polycera conspicua Allan, 1932

Taxonomy: *Polycera capensis* (Fig. 8b) is a distinctive species, but has been described under several names. Specimens from southern Africa are variably colored. The ground color may be white to grey and black pigment may be present or absent from the notum.

Natural History: *Polycera capensis* is generally associated with bryozoans of the genus *Bugula*, on which it feeds. It is found from the intertidal fringe to depths of at least 30 m.

Occurrence and Distribution: Specimens have been found along the southern African coast from Luderitz, in Namibia, to Port Alfred. Outside southern Africa, *Polycera capensis* has been found in Sydney Harbor, Australia.

172. ***Polycera quadrilineata*** (Müller, 1776)

Taxonomy: *Polycera quadrilineata* differs from *P. capensis* by the presence of yellow tubercles around the mid-dorsal region and the sides of its body.

Natural History: This species has been found intertidally on the underside of rocks among arborescent bryozoans.

Occurrence and Distribution: Numerous specimens have been found in Knysna Lagoon and Algoa Bay. Until this species was found in South Africa, it was known only from the northeastern Atlantic and the Mediterranean.

173. ***Polycera hedgpethi*** Marcus, 1964

Taxonomy: A morphological comparison and systematic discussion of this species is provided by Gosliner (1982). It is readily distinguishable from *P. capensis*, *P. quadrilineata*, and *P.* sp. by its green ground color.

Natural History: *Polycera hedgpethi* feeds upon the purple-brown bryozoan, *Bugula neritina*, which is epizootic on eel grass in shallow estuaries.

Occurrence and Distribution: *Polycera hedgpethi* has been found near the mouth of the Keurbooms River, near Plettenberg Bay. Prior to its discovery in South Africa, this species was known only from the Pacific coast of Mexico and California. It has subsequently been recorded from Australia and New Zealand (Willan and Coleman 1984), where it is considered to be introduced. It is unlikely that the species has been introduced to southern Africa.

174. ***Polycera* sp.**

Taxonomy: *Polycera* sp. resembles *P. capensis* and *P. quadrilineata* in its coloration, but differs significantly by having three, rather than one, extra, branchial processes per side. It appears to be undescribed.

Natural History: This species was collected under a rock in 2 m of water.

Occurrence and Distribution: The single specimen was collected from Vetchies Pier, north of the entrance to Durban Harbor.

175. ***Plocamophorus maculatus*** (Pease, 1860)

Taxonomy: *Plocamophorus maculatus* can be distinguished by its pale body color with opaque white markings and yellow-orange spots. *P. apheles* Barnard, 1927, was described from a single specimen collected at Nahoon Estuary near East London. Barnard's description is too superficial to compare *P. apheles* with other species until further specimens have been collected from the type locality.

Natural History: This species was found under a dead coral head at a depth of 15 m.

Occurrence and Distribution: The single specimen was collected at Sodwana Bay National Park. Previously, *P. maculatus* was only known from the type locality, Hawaii.

176. ***Plocamophorus* sp.**

Taxonomy: This species appears to be undescribed and differs from *Plocamophorus maculatus* by its brown color with opaque white spots. *Plocamophorus* sp. has an additional pair of papillae with spherical apices.

Natural History: *Plocamphorus* sp. was found in an intertidal rock pool at night.

Occurrence and Distribution: The single specimen was collected at Jesser Point, Sodwana Bay National Park.

177. ***Kaloplocamus ramosus*** (Cantraine, 1835)

Taxonomy: *Kaloplocamus ramosus* can be recognized by its highly branched papillae. It is allied to *Plocamophorus*, but lacks club shaped papillae near the gills.

Natural History: This species inhabits deep water and is usually dredged or trawled in several hundred meters of water.

Occurrence and Distribution: *Kaloplocamus ramosus* has been found in southern Africa from Cape Point to Transkei. Outside of southern Africa, *K. ramosus* has been found from the Mediterranean and from Australia and Japan.

178. ***Aegires* sp.**

Taxonomy: The genus *Aegires* is characterized by numerous papillae which cover the dorsal surface, some of which are on the rhinophoral sheaths. The species found in South African waters appears to be undescribed.

Natural History: This species has been found in 1 m of water on a vertical rock wall which was encrusted with sponges, hydroids, and tunicates. Although feeding has not been observed, other species of *Aegires* have been observed to feed upon calcareous sponges.

Occurrence and Distribution: Several specimens of *Aegires* sp. have been found from the Atlantic coast of the Cape Peninsula to Port Elizabeth. This species appears to be restricted to South Africa.

179. ***Limacia clavigera*** (Müller, 1776)
Synonym: *Triopa lucida* Stimpson, 1855

Taxonomy: *Limacia clavigera* is recognized by its elongate papillae with orange tips, which resemble the cerata of aeolid nudibranchs. However, the papillae lack extensions of the digestive gland.

Natural History: *Limacia clavigera* feeds on encrusting Bryozoa of the genus *Membranipora*. It is found from the intertidal border to 20 m. The egg mass is a low, flat ribbon, closely appressed to the substrate, consisting of one to two whorls.

Occurrence and Distribution: In South African waters, *Limacia clavigera* has been found from Saldanha Bay to Port Alfred. It is a common inhabitant of kelp bed communities in the Cape Province. *Limacia clavigera* is also know from North Atlantic of Europe and Mediterranean waters, from which it was first described.

180. ***Nembrotha livingstonei*** Allan, 1933

Taxonomy: *Nembrotha livingstonei* is characterized by its reddish brown ground color with red-orange spots. The rhinophores and gills are brownish, ornamented with opaque white. In all aspects of its external anatomy, the specimen from southern Africa resembles Allan's original description.

Natural History: This species was found crawling on the compound tunicate on which it was photographed in 15 m of water.

Occurrence and Distribution: *Nembrotha livingstonei* was collected together with *N. purpureolineata* at La Mercy, Natal. This constitutes the first record of this species since its original description. *Nembrotha livingstonei* was previously known only from a single specimen collected in western Australia.

181. ***Nembrotha purpureolineata*** O'Donoghue, 1924

Taxonomy: The brilliant pattern of coloration readily distinguishes this species from closely related members of the genera *Nembrotha*, *Roboastra*, and *Tambja*. The genera can only be distinguished by their radular morphology. Baba (1976) stated that *N. lineolata* may be synonymous with *N. purpureolineata* O'Donoghue, 1924. However, *N. lineolata* possesses more, thinner, brown longitudinal lines, which are often broken, lacks yellow pigment on the notum, and has yellow pigment on the base of the gill rachis and on the rhinophore sheaths. These differences appear to be consistent and warrant the separation of the species.

Natural History: This species was crawling across a dead coral reef in 11 m of water.

Occurrence and Distribution: The single specimen found at La Mercy, Natal represents the first record of this species from southern Africa. *Nembrotha purpureolineata* is widely distributed in the Indo-Pacific and has been found from western Australia and Japan.

182. ***Tambja capensis*** (Bergh, 1907)

Taxonomy: The uniform black color and green marginal line distinguish *Tambja capensis*. This species, described in the genus *Nembrotha*, has radular characteristics which indicate that it should be placed in *Tambja*. Macnae (1958) stated that *Tambja capensis* should be regarded as synonymous with *Nembrotha morosa* Bergh, 1877. However, there are several radular and reproductive differences, as well as distinguishing features of their coloration. Therefore, they are here regarded as distinct species.

Natural History: *Tambja capensis* feeds on arborescent bryozoans and has been found from intertidal locales to at least 30 m.

Occurrence and Distribution: This species is rare intertidally but is one of the most common nudibranchs in shallow subtidal waters. It has been found on both coasts of the Cape Peninsula and specimens have been found as far east as Port Elizabeth. *Tambja capensis* appears to be endemic to the temperate waters of the Cape Province.

183. ***Tambja morosa*** (Bergh, 1877)

Taxonomy: Several workers, including Barnard (1927) and Macnae (1958), considered *T. capensis* a synonym of *T. morosa*. *T. morosa* can be distinguished by the interrupted, blue-green notal margin and pigment of the same color which extends on to the gills. The two species are geographically isolated, with *T. morosa* occurring in the Indo-Pacific tropics, while *T. capensis* is apparently endemic to the temperate waters of the Cape Province.

Natural History: Specimens were observed crawling in crevices that contained arborescent bryozoans in 1 to 4 m of water.

Occurrence and Distribution: Six specimens, collected at Mbibi, 20 km north of Sodwana Bay National Park, constitute the first record of this species in southern Africa. *T. morosa* has previously been collected from the Philippines and Indonesia.

184. ***Tambja* sp.**

Taxonomy: On the basis of its radula, this species should be placed in *Tambja*. It is distinct from *T. capensis* and *T. morosa* and has several longitudinal green lines along the body and gills. In the others there is a single line along the margin of the notum. *Tambja* sp. appears to be undescribed.

Natural History: *Tambja* sp. has been found under stones in 3 m of water.

Occurrence and Distribution: The single specimen was collected at the mouth of Durban Harbor. It has not been found outside South African waters.

185. ***Roboastra gracilis*** (Bergh, 1877)

Taxonomy: *Roboastra* can be distinguished from *Tambja* and *Nembrotha* on the basis of its radular morphology.

Natural History: *Roboastra gracilis* has been found crawling on rocks in 1 or 2 m of water on tropical reefs.

Occurrence and Distribution: The first record of this species from the Indian Ocean is from specimens collected at Sodwana Bay National Park. Previously, *Roboastra gracilis* has been recorded from the Philippines, Australia, and Japan.

186. ***Roboastra luteolineata*** (Baba, 1936)

Taxonomy: *Roboastra luteolineata* is distinguishable from the preceding species by its larger body size, mid-dorsal lines, which extend on to the gills, and highly branched, rather than simply pinnate gills.

Natural History: *Roboastra luteolineata* has been found together with *R. gracilis* in shallow water on tropical reefs.

Occurrence and Distribution: Both South African specimens were collected at Sodwana Bay National Park. Outside of South Africa, *Roboastra luteolineata* has been found from Japan and Okinawa. It is probably widely distributed throughout the Indo-Pacific.

187. ***Gymnodoris inornata*** (Bergh, 1880)

Taxonomy: This species is characterized by its orange body color and complete circle of gills. The radula closely matches that described by Baba (1949).

Natural History: Two specimens were collected in 2 to 3 m of water.

Occurrence and Distribution: Previously, this species had been recorded from southern Africa, based upon specimens collected by Macnae from Inhaca Island, Mozambique. The present two specimens were found near the mouth of Durban Harbor. Outside southern Africa, this species has been recorded from Japan and Indonesia, although Edmunds (1971) noted that the Indonesian specimen had a different radular morphology.

188. ***Gymnodoris okinawae*** Baba, 1936

Taxonomy: *Gymnodoris okinawae* is distinguishable by its white body color with orange spots or lines and white tubercles. It has a complete circle of gills.

Natural History: This species inhabits intertidal rock pools on tropical reefs. Although no observations of feeding were made by the author, Kay and Young (1969) noted that *Gymnodoris okinawae* feeds on sacoglossan opisthobranchs in Hawaii.

Occurrence and Distribution: The three South African specimens were found at Sodwana Bay National Park. *Gymnodoris okinawae* is widely distributed in the Indo-Pacific and has been recorded from Tanzania, Okinawa, and Hawaii.

189. ***Gymnodoris alba*** (Bergh, 1877)

Taxonomy: Externally, *Gymnodoris alba* closely resembles *Gymnodoris* sp. 2, but differs by having a nearly complete circle of gills and a much broader radula.

Natural History: Specimens were collected under coral rubble in 1 to 2 m of water.

Occurrence and Distribution: Three specimens collected from Vetchies Pier, north of Durban Harbor, represent the first record of this species from southern Africa. Elsewhere, *Gymnodoris alba* is known from Cargados Carajos in the Indian Ocean, the Philippines, Japan, and Hawaii.

190. ***Gymnodoris* sp. 1**

Taxonomy: *Gymnodoris* sp. 1 is, presumably, undescribed. Although it resembles *G. bicolor* (Alder and Hancock, 1864) in external appearance, it differs significantly in its radular morphology and position of the reproductive opening.

Natural History: Specimens were found together with *Gymnodoris okinawae* in tropical intertidal rock pools.

Occurrence and Distribution: The single specimen was found at Sodwana Bay National Park.

191. ***Gymnodoris* sp. 2**

Taxonomy: The blunt head, with poorly developed oral tentacles, distinguishes this species. Edmunds (1971) discussed an undescribed species of *Gymnodoris* from Tanzania which closely resembles *Gymnodoris* sp. 2 in its external and internal morphology. Insufficient material prohibits a more detailed comparison of the present species to that treated by Edmunds.

Natural History: The single specimen was found on a rock in 1 m of water.

Occurrence and Distribution: In South Africa this species is known only from the mouth of Durban Harbor.

192. ***Lecithophorus capensis*** Macnae, 1958

Taxonomy: The genus *Lecithophorus* contains only one described species. *Lecithophorus* differs from the closely related genus *Gymnodoris* in its radular morphology and by possessing a thick notal margin.

Natural History: *Lecithophorus capensis* has been found in the intertidal region to at least 30 m. Although Macnae stated that it probably feeds on colonial tunicates, all specimens observed in this study were feeding on erect Bryozoa.

Occurrence and Distribution: Macnae described this species based on intertidal specimens from False Bay. *Lecithophorus capensis*, however, is uncommon intertidally, but is one of the most common subtidal opisthobranchs on both sides of the Cape Peninsula. This species appears to be endemic to the Cape Province.

193. ***Lecithophorus* sp.**

Taxonomy: *Lecithophorus* sp. differs from *L. capensis* in its coloration and in the morphology of the buccal mass. This species appears to be undescribed.

Natural History: This species inhabits shallow subtidal reefs of about 10 m depth. It is associated with and feeds upon bryozoans.

Occurrence and Distribution: Eight specimens have been collected, all within Algoa Bay. *Lecithophorus* sp. appears to be endemic to southern Africa.

Family Vayssieridae

194. ***Okadaia elegans*** Baba, 1931

Taxonomy: *Okadaia elegans* can be distinguished from other southern African opisthobranchs by its small body size, red-orange coloration, and absence of gills. The relationship of this species to *O. cinnabareus* (Ralph, 1944) from New Zealand needs to be examined in more detail.

Natural History: This species is associated with and feeds upon serpulid polychaete worms of the genus *Spirorbis*. Specimens are found underneath rocks in the mid-intertidal zone.

Occurrence and Distribution: Numerous specimens, collected from Gonubie, near East London, constitute the first record of this species from the Indian Ocean. Outside of southern Africa, *O. elegans* is known from Japan, Enewetak Atoll, and Hawaii.

Suborder Dendronotacea

Family Bornellidae

195. ***Bornella adamsi*** Gray, 1850

Taxonomy: Eight species of *Bornella* have been synonymized under *Bornella adamsi*. The species is characterized by relatively small extra-branchial processes and a white body with orange reticulations.

Natural History: *Bornella adamsi* has been found from the intertidal fringe to 3 m, where it feeds upon a variety of thecate hydroids.

Occurrence and Distribution: *Bornella adamsi* has been found in South Africa from Port St. Johns, Transkei to Durban Harbor. This species is widely distributed throughout the Indo-Pacific, eastward to the Hawaiian Islands.

196. ***Bornella anguilla*** Johnson, 1983

Taxonomy: This species, with its large, paddle-like extra-rhinophoral and extra-branchial processes, is clearly distinct from all other species of the genus.

Natural History: *Bornella anguilla* was found in 18 m of water, crawling over large colonies of sertularid hydroids on which it presumably feeds. When disturbed, *B. anguilla* actively swims by laterally undulating its body. This movement can be sustained for several minutes.

Occurrence and Distribution: A single specimen was found at Sodwana Bay National Park. This recently described species is present at Enewetak Atoll and western Australia, as well (Patrick Baker, personal communication).

Family Scyllaeidae

197. ***Notobryon wardi*** Odhner, 1936

Taxonomy: Thompson and Brown (1981) recently recorded this species (Fig. 7h) from South African waters and noted that previous records erroneously refer to it as *Scyllaea pelagica*.

Natural History: *Notobryon wardi* is found from the intertidal zone to 15 m. It feeds on campanularid hydroids. It is most commonly found intertidally. This species, when disturbed, swims by lateral movements of the body.

Occurrence and Distribution: In South Africa, *Notobryon wardi* has been found from the Atlantic coast at Eland's Bay to the Indian Ocean at Jeffrey's Bay. This species was described from Australia and has also been found in Japan.

Family Tethyidae

198. ***Melibe rosea*** Rang, 1829

Taxonomy: *Melibe rosea* (Fig. 8c) is characterized by its large oral hood and pink to red-orange coloration, often with opaque white encrustations on the notum and cerata.

Natural History: In the rocky intertidal region, *Melibe rosea* is one of the most common nudibranchs around the Cape Province. It is a planktonic feeder and captures minute amphipod crustaceans with the tentacles of its oral hood. The egg mass is a tall collar of a few whorls.

Occurrence and Distribution: *Melibe rosea* has been found along much of the coast of the Cape Province, from Port Nolloth on the Atlantic to Port Alfred along the Indian Ocean coast. It appears to be endemic to the Cape Province.

199. ***Melibe* sp.**

Taxonomy: *Melibe* sp. is distinguishable from *Melibe rosea* by its uniform white color, larger size, and by the presence of an appendage on the posterior side of the rhinophore sheath.

Natural History: Very little natural history data are available for this species. Specimens have been found on subtidal rock outcrops in 10 to 20 m of water. Like the preceding species, *Melibe* sp. feeds on crustaceans, which it captures with its oral hood.

Occurrence and Distribution: Four specimens have been found at Oudekraal on the Atlantic coast of the Cape Peninsula.

Family Dotoidae

200. ***Doto coronata*** (Gmelin, 1791)

Taxonomy: The South African specimens of *Doto coronata* (Fig. 14a) resemble the North Atlantic specimens except that they sometimes lack dark mottling on the notum. These slight differences are considered unimportant at the species level and the South African material is considered to be synonymous with that from the North Atlantic.

Natural History: *Doto coronata* feeds on several species of campanularid hydroids. It lays a convoluted, coiled egg mass, usually along the main axis of hydroid colonies. Specimens have been found from the intertidal zone to a depth of 10 m.

Occurrence and Distribution: South African specimens have been found from the Atlantic coast of the Cape Peninsula to Knysna. They are common on pier pilings and floats, as well as on hydroids, which are epizootic on kelp fronds. *Doto coronata* is known from both coasts of the North Atlantic and from the Mediterranean.

201. ***Doto pinnatifida*** (Montagu, 1804)

Taxonomy: The tubercles on the cerata of *Doto pinnatifida* are more elongate than in *D. coronata.* Barnard (1927) recorded *D. splendida* Trinchese, 1881 from South African waters and noted that it was possibly synonymous with *D. pinnatifida.*

Natural History: Specimens have been found in association with campanularid hydroids to a depth of 30 m. Egg masses are frequently deposited on the hydroid colonies.

Occurrence and Distribution: Specimens have been collected along the Atlantic and Indian Ocean coasts of the Cape Province from Oudekraal to Knysna. Outside of southern Africa, *Doto pinnatifida* is known only from the British Isles and the Atlantic coast of France.

202. ***Doto rosea*** Trinchese, 1881
Synonym: *Doto cinerea* Trinchese, 1881

Taxonomy: This species was recorded from South Africa by Barnard (1927) as *Doto cinerea*. Pruvot-Fol (1954) considered Trinchese's *D. rosea* and *D. cinerea* as synonyms.

Natural History: In South Africa, *Doto rosea* is associated with intertidal campanularid hydroids.

Occurrence and Distribution: Barnard stated that *Doto rosea* was common in False Bay. Only occasional specimens have been found in recent years, also within False Bay, at Clovelly and Dalebrook. Outside southern Africa, *D. rosea* is known from the Mediterranean.

Family Marianinidae

203. ***Marianina rosea*** Pruvot-Fol, 1930
Synonym: *Aranucus bifidus* Odhner, 1936

Taxonomy: The purple coloration with yellowish, bifid cerata and red rhinophores clearly differentiates *Marianina rosea* from other southern African nudibranchs.

Natural History: *Marianina rosea* was collected on arborescent hydroids in 1 to 4 m of water.

Occurrence and Distribution: The three specimens of *Marianina rosea* collected at Mbibi, 20 km north of Sodwana Bay National Park, represent the first records of this species from the Indian Ocean. *Marianina rosea* has previously been recorded from the type locality, New Caledonia, from the Gilbert Islands and Guam.

Family Tritoniidae

204. ***Tritonia nilsodhneri*** Marcus, 1983
Synonym: *Duvaucelia odhneri* Tardy, 1963, non Marcus, 1959

Taxonomy: *Tritonia odhneri* (Tardy, 1963) is a junior homonym of *T. odhneri* Marcus, 1959. For this reason, Marcus (1983) renamed this species.

Natural History: *Tritonia nilsodhneri* preys on gorgonians of the genus *Eunicella*. It is cryptically colored, the animal varying with the coloration of the gorgonians. Its convoluted egg mass, shown in the photo, varies in its coloration also according to the color of the gorgonians. This species is found in depths from 10 to 30 m.

Occurrence and Distribution: In South Africa, *Tritonia nilsodhneri* has been found only along the Atlantic coast of the Cape Peninsula at Oudekraal. *Tritonia nilsodhneri* is known from the Atlantic coast of France, England, and Ireland.

205. ***Tritonia* sp. 1**

Taxonomy: This species of *Tritonia* appears to be undescribed. It is characterized by velar tentacles with elongate bifurcations.

Natural History: *Tritonia* sp. 1 appears to feed exclusively on soft corals of the genus *Parerythropodium.* It is very cryptic when found with its prey and varies in color from white to brown to purple, depending on the form of *Parerythropodium* on which it is feeding. Its egg mass is opaque white and highly convoluted. This species has been found from the intertidal boundary to a depth of 10 m.

Occurrence and Distribution: This species appears to be endemic to southern Africa, having been found from Buffels Bay, in False Bay, to Port Elizabeth.

206. ***Tritonia* sp. 2**

Taxonomy: Like the preceding species, *Tritonia* sp. 2 appears to be undescribed. Both its tubercular velar tentacles and gills are short.

Natural History: *Tritonia* sp. 2 has been found in 30 to 40 m of water. Although little is known of its habits, *Tritonia* sp. 2, like other members of the genus, probably feeds upon octocorals.

Occurrence and Distribution: One specimen has been collected by means of scuba from the Atlantic coast of the Cape Peninsula, off Llandudno. A second specimen was dredged off Jeffrey's Bay. It appears that this species is endemic to South Africa.

207. ***Marionia* sp. 1**

Taxonomy: *Marionia* sp. 1 can be distinguished from all other South African tritoniids by the reticulate pattern of ridges on its notum. This species appears to be undescribed.

Natural History: *Marionia* sp. 1 was collected in a shallow rock pool in the mid-intertidal zone at night.

Occurrence and Distribution: The single specimen was collected at Jesser Point, Sodwana Bay National Park.

208. ***Marionia* sp. 2**

Taxonomy: This species appears to be undescribed. It differs from other southern African tritoniids in its coloration. It is the only species in the region that possesses blue spotting on the notum.

Natural History: The single specimen was found on the surface of a rock in 10 m of water.

Occurrence and Distribution: The only specimen of *Marionia* sp. 2 was collected from Algoa Bay. This species is not known outside of southern Africa.

209. ***Marioniopsis cyanobranchiata*** (Rüppell and Leuckart, 1831)

Synonym: *Marionia arborescens* Bergh, 1890

Taxonomy: *Marioniopsis* has highly branched gills and a larger body than other South African tritoniids. This species was previously recorded from South Africa by Eliot (1910) as *M. arborescens*.

Natural History: *Marioniopsis cyanobranchiata* has been found intertidally, where it appears to feed on the soft coral, *Xenia* sp.

Occurrence and Distribution: Since Eliot (1910) recorded a single specimen from Scottburgh, Natal, two further specimens have been found, one each at Umgazana, Transkei and Park Rynie. *Marioniopsis cyanobranchiata* was described from the Red Sea and is also known from the Philippines and Tanzania.

Suborder Arminacea

Family Arminidae

210. ***Armina gilchristi*** (Bergh, 1907)

Taxonomy: Several species of *Armina* (Fig. 11c) from southern Africa have been described from preserved specimens, and at present, can be distinguished only by their radular morphology. *A. gilchristi* is the first species to be observed alive.

Natural History: Specimens had been found only by dredging in relatively deep water (70 to 80 m). However, one specimen was recently trawled from a depth of 30 m.

Occurrence and Distribution: *Armina gilchristi* appears to be endemic to southern Africa, where it is known from Cape Point to Jeffrey's Bay.

211. ***Dermatobranchus* sp. 1**

Taxonomy: *Dermatobranchus* differs from *Armina* by lacking a distinct ridge of tissue separating the rhinophores from the notum and by an absence of distinct gill ridges between the mantle and foot. This species is undescribed.

Natural History: This species feeds on the globular soft coral, *Sarcophytum trochiforme*, in 10 to 15 m of water.

Occurrence and Distribution: Numerous specimens have been found at Oudekraal along the Atlantic coast of the Cape Peninsula and at Port Elizabeth. *Dermathobranchus* sp. 1 appears to be endemic to South Africa.

212. ***Dermatobranchus* sp. 2**

Taxonomy: Like the preceding species, *Dermatobranchus* sp. 2 appears to be undescribed. It is distinguishable by interrupted, longitudinal ridges on the notum.

Natural History: *Dermatobranchus* sp. 2 has been found in rocky intertidal pools, where it feeds on the soft coral, *Xenia* sp. It is often partially burrowed in sand.

Occurrence and Distribution: Numerous specimens have been found at Umgazana, Transkei; it is probably more widely distributed.

213. ***Dermatobranchus* sp. 3**

Taxonomy: *Dermatobranchus* sp. 3 can be distinguished by its white notum and longitudinal ridges. Brown spots are present on the anterior portion of the notum. Like the preceding two species, *Dermatobranchus* sp. 3 appears to be undescribed.

Natural History: This species is found on dead and living reefs in 1 to 15 m of water. It appears to be associated with soft corals.

Occurrence and Distribution: Seven specimens have been found in and around Sodwana Bay National Park.

214. ***Dermatobranchus* sp. 4**

Taxonomy: This species is similar in appearance to *Dermatobranchus* sp. 1, but has brown mottling and more longitudinal lines on the notum. There are also significant radular differences between the two species. *Dermatobranchus* sp. 4 also is undescribed.

Natural History: *Dermatobranchus* sp. 4 inhabits subtidal reefs at about 20 m depth.

Occurrence and Distribution: The single specimen was collected from Bakoven, on the Atlantic coast of the Cape Peninsula. It has not been found outside southern Africa.

Family Lemindidae

215. ***Leminda millecra*** Griffiths, 1985

Taxonomy: This species has recently been described and belongs to a new genus and family. Externally, it resembles *Telarma*, but differs significantly in its internal morphology.

Natural History: Many specimens have been collected in 10 m of water, but this species appears to be far more common at depths of 30 to 40 m. Their food is unknown.

Occurrence and Distribution: It has been found from Llandudno and Castle Rocks, on both coasts of the Cape Peninsula. Specimens have been collected in 10 m of water from Algoa Bay and deep water specimens have been trawled from the Transkei and Natal coasts. It appears to be endemic to South Africa.

Family Janolidae

216. ***Janolus capensis*** Bergh, 1907

Taxonomy: The Janolidae are related to other arminaceans, but differ in possessing cerata. *Janolus capensis* was originally described from South Africa in 1907. Some records may actually represent *J. longidentatus.*

Natural History: *Janolus capensis* feeds on arborescent bryozoans, especially *Menipea triseriata* and *Onchoporella buskii.* It has been found from the intertidal border to at least 40 m. The egg mass is convoluted and globular, with numerous (30-40) eggs per capsule.

Occurrence and Distribution: *Janolus capensis* is endemic to the Cape Province of South Africa, having been found from Saldanha Bay to East London. It is one of the most common nudibranchs on both sides of the Cape Peninsula.

217. ***Janolus longidentatus*** Gosliner, 1981

Taxonomy: This species is almost identical to *Janolus capensis* in its appearance. It differs by having a rounded, rather than conical, anterior margin of the head. Internally, there are numerous signficant differences between the two species (Gosliner 1981).

Natural History: As in *Janolus capensis*, *J. longidentatus* feeds on the bryozoan *Menipea triseriata*. It is found from the intertidal zone to at least 30 m. The egg mass is a flat spiral, unlike the globular egg mass of *J. capensis*. There are only 5 to 7 eggs per capsule in this species.

Occurrence and Distribution: *Janolus longidentatus* has been found from Cape Point to False Bay. Unlike *Janolus capensis*, it has not been found from the Atlantic coast of the Cape Peninsula. It appears to be endemic to the Cape Province.

218. ***Bonisa nakaza*** Gosliner, 1981

Taxonomy: *Bonisa* differs from *Janolus* in that the cerata lack extensions of the digestive gland. *Bonisa nakaza* is variable in its coloration, and may have blue, salmon, yellow, or orange pigment on the cerata.

Natural History: *Bonisa nakaza* feeds on the heavily calcified bryozoan *Tubucellaria levinseni*. It has been found from just below the low tide mark to at least 30 m in depth. Its egg mass is highly convoluted and is usually attached to branched organisms, such as gorgonians.

Occurrence and Distribution: *Bonisa nakaza* occurs commonly on both sides of the Cape Peninsula and has been found eastwards to Port Elizabeth.

Suborder Aeolidacea

Family Flabellinidae

219. ***Flabellina funeka*** Gosliner and Griffiths, 1981

Taxonomy: *Flabellina funeka* is identifiable by its striking mauve and red coloration and annulate rhinophores (Fig. 7 e). The genus *Flabellina* is large and one of the most morphologically diverse of all the aeolid nudibranch genera. The ceratal arrangement, (Figs. 12 a, 14 d), shape of the rhinophores and elaboration of the anterior margin of the foot (Fig. 13 a) are among the most variable external characteristics.

Natural History: *Flabellina funeka* feeds on species of the athecate hydroid genus *Eudendrium*. Its egg mass is highly convoluted. It has been found in 7 to 17 m of water.

Occurrence and Distribution: *Flabellina funeka* is endemic to the Cape Province, where it appears to be restricted to the Indian Ocean. It has been found from Cape Point and as far east as Port Elizabeth.

220. ***Flabellina capensis*** (Thiele, 1925)

Taxonomy: *Flabellina capensis* is distinguishable by its rugose rhinophores (Fig. 7 g) and the opaque white lines on the notum and cerata. Prior to its rediscovery (Gosliner and Griffiths 1981) this species was known from a single specimen, dredged from Plettenberg Bay. *F. capensis* is similar to, but distinct from, *F. lineata* and *F. browni* (Picton 1980) from European waters.

Natural History: *Flabellina capensis* feeds on the athecate hydroid *Eudendrium* sp. Its egg mass is highly convoluted, consisting of several whorls. This species is commonly found in 10 to 30 m of water.

Occurrence and Distribution: *Flabellina capensis* is known only from South African waters, from Oudekraal, on the Atlantic coast of the Cape Peninsula, to Port Elizabeth. It has recently been rediscovered from the type locality.

221. ***Flabellina*** sp. 1

Taxonomy: *Flabellina* sp. 1 is distinguishable by its papillate rhinophores. It is similar to *F. albomarginata* (Miller, 1971) from New Zealand, but differs externally and in its radular morphology. It is, therefore, considered to be a distinct undescribed species.

Natural History: Like the two preceding species of *Flabellina*, *Flabellina* sp. 1 feeds on hydroids of the genus *Eudendrium*. It has been found at depths from 5 to 15 m.

Occurrence and Distribution: *Flabellina* sp. 1 has only been found along the west coast of False Bay at Castle Rocks and from the Atlantic coast at Oudekraal, but is likely to be more widespread in the Cape Province.

222. ***Flabellina* sp. 2**

Taxonomy: This species appears to be undescribed, but resembles *Flabellina capensis* in having rugose rhinophores. The two differ significantly in their coloration, ceratal arrangement, and internal morphology.

Natural History: *Flabellina* sp. 2 has been found in association with *Eudendrium*, on which it presumably feeds. It has been found at about 10 m depth.

Occurrence and Distribution: This species, like *Flabellina* sp. 1, has been found along the western shore of False Bay, at Windmill Beach and at Dassie Point.

223. ***Flabellina* sp. 3**

Taxonomy: Based upon its coloration, *Flabellina* sp. 3 differs from all described species of the genus. Its groups of cerata are elevated from the notum. It is the only member of the genus from southern Africa which has perfoliate rhinophores.

Natural History: This species has been found in 10 m of water on tropical coral reefs.

Occurrence and Distribution: A single specimen was found at Sodwana Bay National Park. This species also appears to have been found at Enewetak Atoll (Scott Johnson, personal communication).

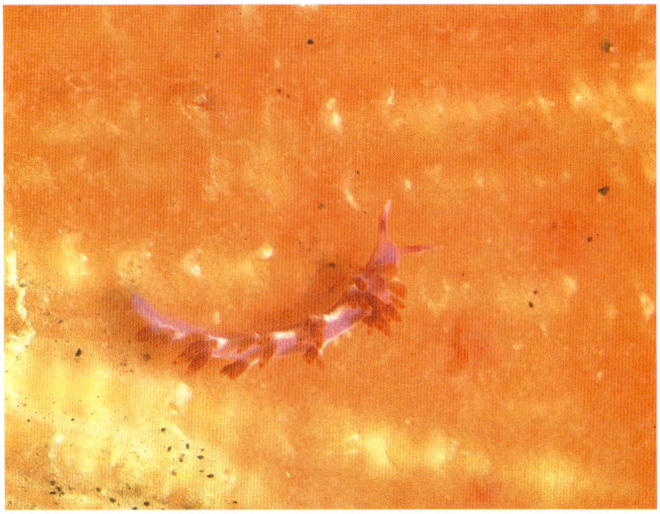

224. ***Coryphellina* sp.**

Taxonomy: *Coryphellina* sp. can be distinguished from other South African aeolids by its papillate rhinophores and elevated ceratal clusters. *C. rubrolineata* O'Donoghue, 1929 and *C. poenicia* (Burn, 1957) are similar to *Coryphellina* sp. in their body form, but differ in their coloration.

Natural History: *Coryphellina* sp. was found crawling on a colony of athecate hydroids in 20 m of water. The egg mass is a flat spiral of several whorls.

Occurrence and Distribution: A single specimen was collected from Aliwal Shoals, off Scottburgh, Natal.

Family Eubranchidae

225. ***Eubranchus* sp. 1**

Taxonomy: The orange and opaque white coloration and inflated cerata distinguish this species from other South African aeolids. No members of the genus have been previously recorded from southern Africa. Most species are small and inconspicuous. This species appears to be undescribed.

Natural History: *Eubranchus* sp. 1 feeds on campanularid hydroids, including *Obelia geniculata*. It is found in shallow subtidal waters from 5 to 10 m depth.

Occurrence and Distribution: This species appears to be endemic to southern Africa. It has been found at Bloubergstrand on the Atlantic coast and at Castle Rocks and Windmill Beach in False Bay.

226. ***Eubranchus* sp. 2**

Taxonomy: Like the preceding species, *Eubranchus* sp. 2 appears to be undescribed. It is distinguishable by its cylindrical cerata (Fig. 14b) and brown spotting on the notum.

Natural History: *Eubranchus* sp. 2 is associated with, and feeds upon, the thecate hydroids *Nemertesia cymodocea* and *Plumularia* sp. Its egg mass is a cylindrical coil of one to one and a half whorls. This species has been found at 10 m.

Occurrence and Distribution: *Eubranchus* sp. 2 appears to be endemic to the Cape Province. It has been collected from Windmill Beach, on the western shore of False Bay, to Knysna.

227. ***Eubranchus* sp. 3**

Taxonomy: *Eubranchus* sp. 3 is characterized by its brown notum with scattered white spots. Its oral tentacles and rhinophores possess a brown ring just below the apex. It appears to be undescribed.

Natural History: *Eubranchus* sp. 3 feeds on campanularid hydroids that grow on the underside of floats.

Occurrence and Distribution: Like the preceding members of the genus, *Eubranchus* sp. 3 is endemic to southern Africa. It has been found only at the mouth of Knysna Lagoon.

Family Embletoniidae

228. ***Embletonia gracilis*** Risbec, 1928

Taxonomy: *Embletonia* is a distinctive genus among the nudibranchs. The oral tentacles have been modified into a rounded velum. The taxonomy of this species was discussed by Gosliner and Griffiths (1981).

Natural History: *Embletonia gracilis* has been found in the intertidal zone, where it feeds on campanularid hydroids, under rocks. It is the only known southern African opisthobranch in which the egg develops directly into a juvenile, bypassing a planktonic larval stage.

Occurrence and Distribution: Specimens have been found in southern Africa along the western shore of False Bay. *Embletonia* gracilis was previously known only from Japan, New Caledonia, Australia, and Hawaii.

Family Tergipedidae

229. ***Cuthona speciosa*** (Macnae, 1954)

Taxonomy: *Cuthona speciosa* is recognizable by its smooth rhinophores, yellow-orange body, and blue or lilac cerata (Figs. 12b, 13a). The variability of this species and its taxonomic position were discussed by Gosliner and Griffiths (1981).

Natural History: *Cuthona speciosa* has been found from the lower intertidal fringe to a depth of at least 30 m, where it feeds on thecate hydroids of the genus *Sertularella*.

Occurrence and Distribution: *Cuthona speciosa* has been found from Oudekraal, on the Atlantic coast of the Cape Peninsula, to Port Elizabeth on the Cape South Coast. It appears to be endemic to southern Africa.

230. ***Cuthona ornata*** Baba, 1937

Taxonomy: *Cuthona ornata* is similar in appearance to *C. speciosa* but differs in having orange pigment on the rhinophores and oral tentacles and in its radular morphology.

Natural History: Specimens were collected on thecate hydroids in 11 m of water.

Occurrence and Distribution: The three specimens collected at La Mercy, Natal represent the first record of this species from southern Africa. *Cuthona ornata* was described from Japan and has subsequently been recorded from Tanzania.

231. ***Cuthona kanga*** (Edmunds, 1970)

Taxonomy: *Cuthona kanga* can be distinguished by its beautiful cerata with gold,, blue, green, and opaque white pigment. The opaque white, transverse band in front of the rhinophores is also characteristic.

Natural History: This species feeds on the thecate hydroid *Thecocarpus* sp. and has been found in depths of 2 to 4 m on dead reefs.

Occurrence and Distribution: Two specimens found at Sodwana Bay National Park represent the first records of this species since its original description. *Cuthona kanga* was described from a single specimen collected in Tanzania.

232. ***Cuthona annulata*** (Baba, 1949)

Taxonomy: *Cuthona annulata* and *C. pinnifera* are the only known members of the genus having annulate rhinophores. Its body is pale white, with cerata that curve toward the center of the body.

Natural History: *Cuthona annulata* was found in a shallow rock pool in the mid-intertidal zone at night.

Occurrence and Distribution: The single specimen found at Jesser Point, Sodwana Bay, is the first record of this species since it was originally described. *Cuthona annulata* was previously known only from Sagami Bay, Japan.

233. ***Cuthona* sp. 1**

Taxonomy: Clearly recognizable by its translucent white color with opaque white markings, this species appears to be undescribed.

Natural History: *Cuthona* sp. 1 feeds upon campanularid hydroids, which grow under rocks in the intertidal region and on pier pilings.

Occurrence and Distribution: It appears that *Cuthona* sp. 1 is restricted to southern Africa, where many specimens have been found on both sides of the Cape Peninsula, at Hout Bay on the Atlantic coast and at St. James in False Bay.

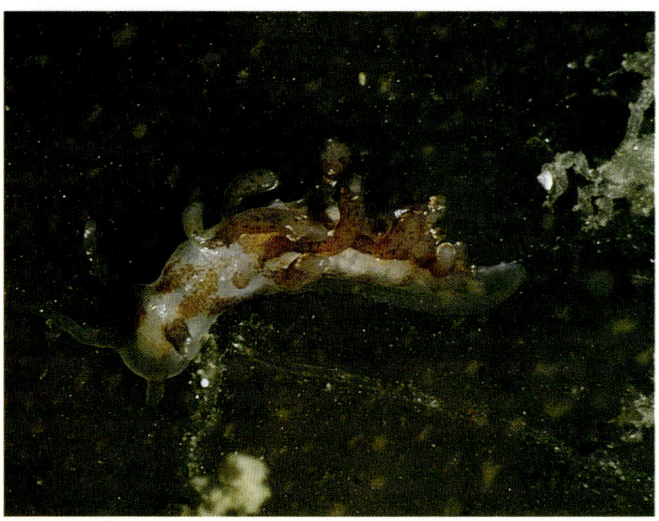

234. ***Cuthona* sp. 2**

Taxonomy: Like the preceding species, this tergipedid is undescribed. Its orange body, with opaque white markings between and posterior to the rhinophores, is distinctive. It could possibly be confused with *Tergipes tergipes*, but they differ in coloration; *Cuthona* sp. 2 (Fig. 13c) has two cerata per row, while *Tergipes* has a single ceras per digestive branch.

Natural History: *Cuthona* sp. 2 feeds on campanularid hydroids that grow on fronds of the kelp *Ecklonia maxima*. It has been found from the water's surface to 20 m depth.

Occurrence and Distribution: *Cuthona* sp. 2 has been found from the Atlantic coast at Bloubergstrand and from the western shore of False Bay at Windmill Beach, Miller's Point, and Castle Rocks. It appears to be endemic to southern Africa.

235. ***Cuthona* sp. 3**

Taxonomy: This striking species is unique within *Cuthona* in that the cerata insert on elevated cushions of tissue as in *Flabellina* funeka and *Flabellina* sp. 3. It appears to be undescribed.

Natural History: *Cuthona* sp. 3 feeds on arborescent thecate hydroids. It has been found in 15 m of water.

Occurrence and Distribution: Only two specimens of this species have been found, both collected from Sodwana Bay National Park. Possibly, it is more widespread in the Indo-Pacific tropics.

236. ***Cuthona* sp. 4**

Taxonomy: The coloration of this minute species of *Cuthona* appears to be distinct and it is likely that it represents an undescribed species. Additional specimens are needed to confirm its taxonomic status.

Natural History: *Cuthona* sp. 4 was found on a shallow coral reef in 1 m of water. It was not found in association with a prey species.

Occurrence and Distribution: The single specimen was collected at Sodwana Bay National Park.

237. ***Cuthona* sp. 5**

Taxonomy: This undescribed species is distinguishable by the presence of brown digestive gland tissue in the anterior portion of the head. The flattened cerata are brown, each with a yellow apex.

Natural History: *Cuthona* sp. 5 was found crawling on a colony of thecate hydroids in 20 m of water.

Occurrence and Distribution: The single specimen was collected at Aliwal Shoals, off Scottburgh, Natal.

238. ***Tergipes tergipes*** Forskäl, 1779

Taxonomy: The South African specimens (Fig. 6d) closely resemble living material examined by the author from the northwestern Atlantic. Some South African specimens have somewhat more elongate cerata than specimens from the North Atlantic, but agree in all other aspects of their morphology.

Natural History: *Tergipes tergipes* feeds on campanularid hydroids which grow on the fronds of the kelp *Ecklonia maxima*. It has been found together with *Cuthona* sp. 2 and *Eubranchus* sp. 1, but is less common. Its egg mass is small and spherical. Specimens have been collected from just below the surface to depths of 10 m.

Occurrence and Distribution: In South African waters, *Tergipes tergipes* has been found on both sides of the Cape Peninsula, at Bloubergstrand and Oudekraal on the Atlantic coast and at Windmill Beach in False Bay. *Tergipes tergipes* is widespread in the Atlantic, having been found from northern Europe, the Mediterranean, the northeastern United States, and Brazil.

239. ***Catriona casha*** Gosliner and Griffiths, 1981

Taxonomy: This recently described species is characterized by its red-orange cerata with opaque white tips, which generally conceal the notum.

Natural History: *Catriona casha* feeds on athecate hydroids of the genus *Tubularia*. It has been found on floats and pier pilings.

Occurrence and Distribution: *Catriona casha* appears to be endemic to South Africa. Since its description from Cape Town Harbor, a few specimens have been found farther north, along the Atlantic coast at Langebaan Langoon, and in the Indian Ocean at Knysna.

240. ***Catriona columbiana*** (O'Donoghue, 1922)

Taxonomy: The status of this species was discussed by Gosliner and Griffiths (1981). It differs from *Catriona casha* by its paler color and more elongate cerata.

Natural History: Like the preceding species, and all other known members of the genus, *Catriona columbiana* probably feeds on hydroids of the genus *Tubularia*.

Occurrence and Distribution: Four specimens of this species have been found in South African waters, all from Cape Town Harbor in Table Bay. *Catriona columbiana* is known from the Pacific coast of North America and Japan. It was probably introduced to South Africa on the hulls of ships.

241. ***Catriona* sp.**

Taxonomy: *Catriona* sp. is presently undescribed and can be distinguished by its cerata, each with a subapical orange band, and by its radular morphology.

Natural History: Specimens were collected in shallow rock pools in the mid-intertidal region, at night. They probably feed on the hydroid *Tubularia* sp., which was present in these rock pools.

Occurrence and Distribution: Seven specimens were collected from Jesser Point, Sodwana Bay National Park.

242. ***Phestilla melanobrachia*** Bergh, 1874

Taxonomy: The taxonomy and morphology of this species have been recently studied by Gosliner (1980) and Rudman (1981*a*). Its coloration varies significantly, depending on its prey (Harris 1968).

Natural History: *Phestilla melanobrachia* has been shown to feed exclusively on corals of the family Dendrophylliidae. In South Africa, it has been found on the dendrophyllid coral, *Tubastrea aurea*, in 2 m of water.

Occurrence and Distribution: Only a single specimen has been found in South African waters; this constitutes the first record of the species from the western Indian Ocean. *Phestilla melanobrachia* is widespread in the tropics of the Indo-Pacific and has been previously recorded from the Philippines, Japan, Hawaii, Australia, Singapore, and Okinawa.

Family Fionidae

243. ***Fiona pinnata*** (Eschscholtz, 1831)

Taxonomy: *Fiona pinnata* is unique among known aeolid nudibranchs in that the posterior surface of each ceras bears an undulating, sail-like membrane (Fig. 14c).

Natural History: *Fiona pinnata* inhabits open oceanic waters where it is found feeding on blue bottles (Portuguese man-of-war) and goose-neck barnacles (*Lepas* sp.). In South Africa, it has been observed feeding on *Lepas* sp. Specimens are often found on logs washed up on the beach, particularly when the logs have a dense covering of barnacles.

Occurrence and Distribution: The only specimen recorded from South Africa was recently found on a log at Witsand, along the southern Cape coast. *Fiona pinnata* is found in all tropical and temperate seas.

Family Facelinidae

244. ***Facelina olivacea*** Macnae, 1954

Taxonomy: *Facelina olivacea* is distinguishable by its annulate rhinophores, green or brown cerata, and opaque white spotting on the notum.

Natural History: This species appears to feed exclusively on athecate hydroids of the genus *Tubularia*. It is most commonly found on floats or pier pilings, but has also been dredged from a depth of 40 m. *Facelina olivacea* has a highly convoluted egg mass of several whorls.

Occurrence and Distribution: Specimens have been found at Saldanha Bay, Langebaan Lagoon, and Cape Town Harbor, all on the Atlantic coast, and from Knysna and Jeffrey's Bay, in the Indian Ocean. This species appears to be endemic to the Cape Province.

245. ***Facelina*** **sp.**

Taxonomy: This species appears to be undescribed and is characterized by annulate rhinophores and orange markings on the notum and oral tentacles.

Natural History: Specimens of *Facelina* sp. have been collected on dead limestone reefs in 1 to 2 m of water.

Occurrence and Distribution: Two specimens have been collected in the vicinity of Sodwana Bay National Park, at Mbibi and Adlam's Reef.

246. ***Pruvotfolia pselliotes*** (Labbé, 1923)
Synonym: *Facelina faurei* Barnard, 1927

Taxonomy: This species was described from southern Africa as a species of *Facelina*. Tardy (1970) has shown that this species should be placed in *Pruvotfolia* owing to its unusual reproductive morphology. More recenty study demonstrates that *P. faurei* and *P. pselliotes* are probably the same species.

Natural History: This species is found on reef flats in about 10 m of water. It has not been found in association with any prey.

Occurrence and Distribution: *Pruvotfolia pselliotes* was known in southern Africa only from Barnard's original description of *Facelina faurei* from Saldanha Bay. Recently, four specimens were collected from Algoa Bay on the Indian Ocean. Outside of southern Africa, this species is known only from France.

247. ***Moridilla brockii*** Bergh, 1888

Taxonomy: *Moridilla brockii* is recognizable by its papillate rhinophores. The cerata are characteristic since some of the inner ones are usually curled and are erected when the animal is disturbed. The shape of the radular teeth and position of the anus are unique to this species.

Natural History: Specimens were collected under an overhanging rock in less than 1 m of water.

Occurrence and Distribution: The two South African specimens were collected from Mbibi, 20 km north of Sodwana Bay National Park. *Moridilla brockii* was originally described from the Sunda Sea and was not recorded again until 1965, when it was redescribed from India (Rao 1965).

248. ***Caloria indica*** (Bergh, 1896)

Taxonomy: *Caloria* closely resembles *Facelina* in its external morphology, but differs in aspects of its reproductive morphology. Confusion regarding *C. indica* and two other species was recently resolved by Rudman (1980). *C. indica* is characterized by smooth rhinophores and an orange head with two opaque white lines.

Natural History: Specimens were collected in rock pools at low tide. They were not associated with any specific prey.

Occurrence and Distribution: Four specimens, collected from Sodwana Bay National Park, constitute the first record of this species from southern Africa. *Caloria indica* is widespread in the Indo-Pacific, having been recorded from Tanzania, Australia, Japan, Indonesia, and Hawaii.

249. ***Caloria* sp. 1**

Taxonomy: *Caloria* sp. 1 appears to represent an undescribed species. It is characterized by perfoliate rhinophores and by a black spot located below the tip of each brick-red ceras. The European *Caloria elegans* (Alder and Hancock, 1845) also has red cerata with black subapical spots, but it has smooth rhinophores.

Natural History: *Caloria* sp. 1 feeds on the thecate hydroid *Campanularia integra*. Its egg mass is cream to yellowish in color and is highly convoluted. It has been found in 5 to 15 m of water.

Occurrence and Distribution: *Caloria* sp. 1 appears to be endemic to the Cape Province. Several specimens have been collected on both sides of the Cape Peninsula, at Oudekraal and Windmill Beach.

250. ***Caloria* sp. 2**

Taxonomy: This species is similar to *Caloria* sp. 1, but differs in its coloration and by its more numerous perfoliations of the rhinophores. It appears to be undescribed.

Natural History: *Caloria* sp. 2 feeds on thecate hydroids and has been collected at 10 to 15 m depth.

Occurrence and Distribution: Unlike *Caloria* sp. 1, which has been collected on both sides of the Cape Peninsula, the several specimens of *Caloria* sp. 2 have been collected only from the Atlantic at Oudekraal. Here, it has often been found together with *Caloria* sp. 1.

251. ***Caloria* sp. 3**

Taxonomy: Like the preceding two members of the genus, this species appears to be undescribed. It is distinguishable by its smooth rhinophores and its cerata, which are opaque white on their outer half.

Natural History: *Caloria* sp. 3 has been found among plants of the siphonaceous green alga *Caulerpa racemosa.*

Occurrence and Distribution: Only one specimen of this species has been collected. It was found at Park Rynie, on the Natal South Coast. This species may be more widely distributed in the Indo-Pacific.

252. ***Favorinus japonicus*** Baba, 1949

Taxonomy: *Favorinus japonicus* is characterized by its nodular cerata and its rhinophores with three distinct swellings. There are usually opaque white markings present on the notum, between the rhinophores and cerata.

Natural History: Specimens were found in rock pools in the mid-intertidal zone. The animals were feeding on the egg masses of the aeolid *Baeolidia palythoae.*

Occurrence and Distribution: Several specimens were collected at Sodwana Bay National Park. Until recently, this species was known only from Japan. Rudman (1980) collected *Favorinus japonicus* from Tanzania and Gosliner (1980) recorded the species from Hawaii. It, therefore, appears to be widely distributed in the tropical Indo-Pacific.

253. ***Favorinus ghanensis*** Edmunds, 1968

Taxonomy: *Favorinus ghanensis* is characterized by its smooth cerata, radular teeth with denticles, and its small penial stylet. It differs externally from *F. japonicus* by its more elongate rhinophores (Fig. 7i), smooth cerata, and the absence of opaque white pigment on the notum.

Natural History: This species has been found under rocks in the mid-intertidal region. In the laboratory, it readily fed on the eggs of *Hypselodoris capensis*, as shown in the figure. Its own egg mass is a tight spiral of three to four whorls.

Occurrence and Distribution: Only a single specimen has been collected in South African from the Heads at the mouth of Knysna Lagoon. Outside of South Africa, *Favorinus ghanensis* is known only from the type locality, Tema, Ghana.

254. ***Godiva quadricolor*** (Barnard, 1927)

Taxonomy: Animals, misidentified as *Godiva quadricolor* from Hawaii (Gosliner 1980), in fact, represent a distinct, undescribed species that is consistently different from the South African specimens in coloration and radular morphology. *G. quadricolor*, with its brown cerata with blue, orange, and yellow pigment, is one of the most striking South African aeolids.

Natural History: *Godiva quadricolor* has been found from rocky intertidal communities to a depth of 15 m. It is a voracious predator, readily consuming other nudibranch species.

Occurrence and Distribution: *Godiva quadricolor* is widespread in the Cape Province, having been found from Bloubergstrand on the Atlantic coast to Richmond, near East London. It is particularly common on the Cape South coast, where it has been collected from Knysna, Tsitsikama National Park, and Port Elizabeth. Outside of southern Africa this species has been reported from Ghana, in West Africa.

255. ***Phyllodesmium serratum*** (Baba, 1949)
Synonym: *Favorinus horridus* Macnae, 1954

Taxonomy: Baba (1955) stated that Macnae's *Favorinus horridus* was synonymous with *Hervia serrata*. Rudman (1981*b*) considered them to be distinct species and placed them within the genus *Phyllodesmium*, since *Hervia* had been discarded. He stated that *P. serratum* has longer cerata than *P. horridus*. Because this character is quite variable, and there are no apparent internal differences, they are, here, considered synonymous, following Baba.

Natural History: In South Africa, *Phyllodesmium serratum* feeds on gorgonians of the genus *Melitodes*. It has been found from the lower intertidal reaches to at least 30 m.

Occurrence and Distribution: Within South Africa this species is widespread, having been collected from Clovelly and Castle Rocks in False Bay to Sodwana Bay. Outside southern Africa, *Phyllodesmium serratum* is known from Japan and Australia.

256. ***Phyllodesmium hyalinum*** Ehrenberg, 1831
Synonym: *Phyllodesmium xeniae* Gohar and Aboul-Ela, 1957

Taxonomy: Species of *Phyllodesmium* have recently been reviewed by Rudman (1981 *b*). *P. hyalinum* can be distinguished by its anal position, unique jaws, and radular morphology.

Natural History: *Phyllodesmium hyalinum* feeds exclusively on soft corals of the genus *Xenia*. In South Africa, it was collected at night in an intertidal rock pool which contained colonies of *Xenia*.

Occurrence and Distribution: The single specimen collected at Jesser Point, Sodwana Bay National Park, represents the first record of this species from southern Africa. *Phyllodesmium hyalinum* is widespread in the Indo-Pacific, where it has been recorded from Tanzania, the Red Sea, the Philippines, and Japan.

257. ***Amanda armata*** Macnae, 1954

Taxonomy: *Amanda armata* is the only known member of its genus. It is characterized by annulate rhinophores with three or four rings and by brown cerata with large opaque white spots.

Natural History: Little is known about the predatory habits of this species. Specimens have been found from the intertidal shallows to depths of 15 m.

Occurrence and Distribution: *Amanda armata* appears to be endemic to southern Africa, having been found from both sides of the Cape Peninsula at Oudekraal and Strandfontein. It is an uncommon species; only four specimens have been collected in 3 years of study.

Family Cratenidae

258. ***Cratena capensis*** Barnard, 1927

Taxonomy: Most members of the genus *Cratena* possess smooth rhinophores and often have red or orange patches on the side of the head. *C. capensis* is unique among described species in having angular, rather than tentacular, foot corners.

Natural History: *Cratena capensis* feeds on a wide variety of hydroids, including *Tubularia solitaria*, *Eudendrium* spp., *Sertularia* sp., and *Obelia geniculata*. Depending on its prey, the coloration of the digestive gland within the cerata varies considerably, from red to dark brown. Its egg mass is a broad coil of one or two whorls. *Cratena capensis* has been found from the intertidal zone to a depth of at least 30 m.

Occurrence and Distribution: *Cratena capensis* is one of the most common aeolid nudibranchs in the Cape Province, to which it appears to be endemic. It has been found from Saldanha Bay to Kleinmond, near Port Alfred.

259. ***Cratena simba*** Edmunds, 1970

Taxonomy: The smooth rhinophores and opaque white markings, surrounded by orange lines, distinguish *Cratena simba* from other South African aeolids.

Natural History: Specimens were found at night in shallow mid-intertidal rock pools.

Occurrence and Distribution: Four specimens collected at Jesser Point, Sodwana Bay National Park, constitute the first record of this species since its original description. *Cratena simba* was previously known only from the type locality, Tanzania.

260. ***Cratena* sp. 1**

Taxonomy: *Cratena* sp. 1 differs from *C. capensis* in several regards. The rhinophores, oral tentacles and cerata are proportionately larger. The foot corners are long and tentacular. The cerata possess a black gland below their apices. These differences also separate *Cratena* sp. 1 from all described species.

Natural History: *Cratena* sp. 1 as been found on pilings in both estuarine and marine waters. It was collected in the vicinity of hydroids, but no food species have been established.

Occurrence and Distribution: Only four specimens of this undescribed species have been found. It has been collected from the railway bridge in Knysna Lagoon and from Plettenberg Bay.

261. ***Cratena* sp. 2**

Taxonomy: *Cratena* sp. 2 appears to be undescribed. Like *C.* sp. 1, it has long tentacular foot corners, but it lacks the subapical black spot within the cerata, which was present in all specimens of *Cratena* sp. 1.

Natural History: *Cratena* sp. 2 was found among hydroids, at 15 m depth, on a living coral reef.

Occurrence and Distribution: The single specimen was collected from Aliwal Shoals, off Scottburgh, Natal.

262. ***Cratena* sp. 3**

Taxonomy: *Cratena* sp. 3 appears to be undescribed and can be distinguished from other South African *Cratena* species by its annulate rhinophores. *C. affinis* (Baba, 1949) also has rhinophores with two annuli, but it possesses tentacular, rather than rounded, foot corners.

Natural History: *Cratena* sp. 3 was found at night in a rock pool in the mid-intertidal region.

Occurrence and Distribution: The single specimen was collected at Jesser Point, Sodwana Bay National Park.

263. ***Cratena* sp. 4**

Taxonomy: This species of *Cratena* most closely resembles *Cratena* sp. 2, which also has tentacular foot corners and lacks a subapical black spot. It differs, however, in possessing an orange ceratal apex rather than a white one. Internal differences also separate the two species.

Natural History: This species is found on shallow rocky reefs in about 10 m of water. No association with food was observed.

Occurrence and Distribution: Two specimens of this species have been collected, both from Algoa Bay. *Cratena* sp. 4 has not been recorded outside southern Africa.

Family Glaucidae

264. ***Glaucus atlanticus*** Forster, 1777

Taxonomy: The taxonomy of this unusually shaped species has been reviewed by Thompson and McFarlane (1967). They concluded that all described members of this genus represent variations within a single species. Its blue and silver coloration and flattened ceratal clusters distinguish this species.

Natural History: *Glaucus atlanticus* lives in the open ocean, where it hangs upside down on the surface film. It maintains buoyancy by swallowing an air bubble. It feeds upon the Portuguese-Man-of-war, *Physalia*, and its relatives, *Velella* and *Porpita*. It is able to incorporate the stinging nematocysts of its prey, which can be used for its own defense. *Glaucus* is frequently washed ashore, onto beaches and into rock pools.

Occurrence and Distribution: Specimens have been found along most of the Indian Ocean coast of South Africa, from False Bay to Natal. Occasional specimens have been found from the Atlantic coast of the Cape Peninsula. This species is circumtropical.

Family Aeolidiidae

265. ***Aeolidiella indica*** Bergh, 1888
Synonyms: *Aeolidiella saldanhensis* Barnard, 1927
Aeolidiella multicolor Macnae, 1954

Taxonomy: The orange markings on the notum and the short foot corners (Fig. 13d) are distinctive. Recently, Gosliner and Griffiths (1981) reviewed the systematics of the *Aeolidiella* species and concluded that the two species named from South Africa, along with four additional species, should all be considered as synonyms of *A. indica.*

Natural History: In southern Africa, *Aeolidiella indica* feeds on the sea anemone *Anthothoe stimpsoni.* It is found commonly in intertidal areas, but extends to waters of at least 10 m depth. The egg mass is a simple spiral, consisting of about three complete whorls.

Occurrence and Distribution: Along the coast of southern Africa, *Aeolidiella indica* has been found from Saldanha Bay on the Atlantic coast to the tropical waters of the Natal South coast at Ramsgate. The species is circumtropical.

266. ***Aeolidiella alba*** Risbec, 1928

Taxonomy: *Aeolidiella alba* is superficially similar to *Favorinus japonicus* and *F. ghanensis* by virtue of its rhinophores with swollen knobs along their length. It can be distinguished from *Favorinus* by its rounded foot corners.

Natural History: This species has been observed to feed on small acontiate white sea anemones. It inhabits tidal swimming pools and mid-intertidal rock pools.

Occurrence and Distribution: Along the Natal coast, several specimens of *Aeolidiella alba* have been found at Salt Rock and at Sodwana Bay National Park. *Aeolidiella alba* is widespread throughout the Indo-Pacific, where it has been recorded from Tanzania, New Caledonia, Australia, and Mexico. Recently, it has been recorded from the Caribbean.

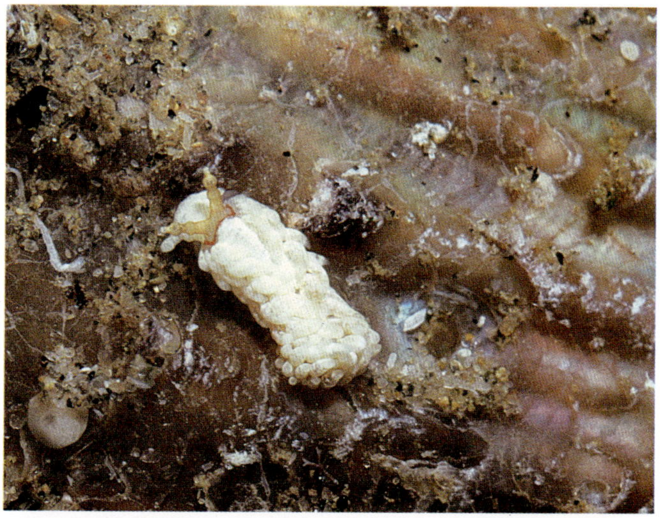

267. ***Berghia chaka*** Gosliner, 1985

Taxonomy: *Berghia chaka* differs from all described members of the genus by having rounded foot corners. It differs from *Baeolidia palythoae* by having most of its cerata arranged in arches, rather than in linear rows.

Natural History: Specimens were collected at night in rock pools.

Occurrence and Distribution: Two specimens of *Berghia chaka* have been collected from Jesser Point, Sodwana Bay National Park. This species is known only from southern Africa.

268. ***Baeolidia palythoae*** Gosliner, 1985

Taxonomy: This recently described species can be distinguished by its papillate rhinophores (Fig. 7f), rounded foot corners and flattened, leaf-like cerata. It is closely allied to *Baeolidia harrietae*.

Natural History: *Baeolidia palythoae* feeds exclusively on zoanthid soft corals, preferring *Palythoa nelliae*. Its egg mass is a cream-yellow, undulating spiral, which is generally laid on the sides of zoanthid polyps.

Occurrence and Distribution: Numerous specimens have been found along the Natal coast, from Umgazana, Transkei to Sodwana Bay National Park. *Baeolidia palythoae* has also been recorded from the Seychelles.

References

Baba, K.

1949. *Opisthobranchia of Sagami Bay, Collected by his Majesty The Emperor of Japan.* Tokyo: Iwanami Shoten.

1955. *Opisthobranchia of Sagami Bay. Supplement.* Tokyo: Iwanami Shoten.

1974. Some comments on *Lobiger souverbiei.* Fischer, 1856, reidentified, from Japan. *Veliger 16*: 253-257.

1976. Two species of *Nembrotha* (*s.s.*) new to Japan (Nudibranchia:Doridoidea:Polyceridae). *Veliger* **19**:131-135.

Barnard, K.H.

1927. South African nudibranch Mollusca, with descriptions of new species, and a note on some specimens from Tristan d'Acunha. *Ann. S. Afr. Mus.* **25:** 171-215.

1932. Note on the true genus of *Doris* (?) *kalkensis* Barnard. *J. Conch. London.* **19**: 256-257.

1933. Description of a new species of *Thecacera. J. Conch. London.* **19:** 294-295.

1934. A new species of *Goniodoris* from South Africa. *J. Conch. London.* **20**: 2-3.

1963*a*. Deep sea Mollusca from west of Cape Point, South Africa. *Ann. S. Afr. Mus.* **46**: 407-452.

1963*b*. Contributions to the knowledge of South African Marine Mollusca. Part IV. Gastropoda:Prosobranchiata: Rhipidoglossa, Docoglossa. Tectibranchiata. Polyphacophora. Solenogastres. Scaphopoda. *Ann. S. Afr. Mus.* **47**: 201-360.

Bartsch, P.

1915. Report on the Turton collection of South African marine mollusks, with additional notes on other South African shells contained in the United States National Museum. *Bull. U.S. Nat. Mus.* **91**: 1-305.

Bergh, R.

1866. Bidrag til en Monographi af Pleurophyllidierne, en Familie af de Gastraeopode Mollusker. *Naturh. Tidsskr. (3)* **4**: 1-81.

1888. *In: Malacologische Untersuchungen.* Sect. 2, **3**: 755-814.

1907. The Opisthobranchiata of South Africa. *Trans. S. Afr. Phil. Soc.* **17**: 1-144.

Burn, R.

1974. The taxonomy and distribution of *Marianina rosea* (Pruvot-Fol, 1930) and *Thecacera pacifica* (Bergh, 1883) *comb. nov.* (Opisthobranchia:Dendrorotacea and Doridacea). *Veliger* **16**: 305-306.

Edmunds M.

1971. Opisthobranchiate Mollusca from Tanzania (suborder: Doridacea). *Zool. J. Linn. Soc.* **50**: 339-396.

Edmunds, M., and T. Thompson.

1972. Opisthobranciate Mollusca from Tanzania IV. Pleurobranchomorpha, Dendronotoidea and Arminoidea. *Proc. Malac. Soc. Lond.* **40:** 219-234.

Eliot, C.

1903. Notes on some new or little known members of the family Doridiidae. *Proc. Malac. Soc. London.* **5**: 331-337.

1904. On some nudibranchs from East Africa and Zanzibar. Part 4. *Proc. Zool. Soc. London.* **1**: 380-406.

1905. On some nudibranchs from the Pacific, including a new genus, *Chromodoridella. Proc. Malac. Soc. London.* **6**: 229-238.

1910. On some nudibranchs from the coast of Natal. *Ann. Natal Mus.* **2**: 221-225.

Engel, H.

1927. Drei neue Arten der Gattung *Aclesia* (Rang) Bergh, 1902.. *Zool Anz.* **69**: 180-187.

Ghiselin, M. T.

1966. Reproductive function and the phylogeny of opisthobranch gastropods. *Malacologia* **3**: 327-378.

Gilchrist, J.

1900. The genus *Paraplysia* with description of a new species. *Trans. S. Afr. Phil. Soc.* **11**: 121-124.

Gosliner, T.

1980. The systematics of the Aeolidacea (Nudibranchia: Mollusca) of the Hawaiian Islands with the descriptions of two new species. *Pacif. Sci.* **31**: 37-77.

1981*a*. Origins and relationships of primitive members of the Opisthobranchia (Mollusca: Gastropoda). *Biol. J. Linn. Soc.* **16**: 197-227.

1981*b*. The South African Janolidae (Mollusca, Nudibranchia) with the description of a new genus and two new species. *Ann. S. Afr. Mus.* **86**: 1-42.

1982. A new record of the nudibranch gastropod *Polycera hedgpethi*, Marcus from the Indian Ocean of South Africa. *J. Moll. Stud.* **48**: 30-35.

1984. Two new species of *Gastropteron* (Gastropoda: Opisthobranchia) from Southern Africa. *Publ. Seto Mar. Biol. Lab.* **29**: 231-247.

1985. The aeolid nudibranch family Aeolidiidae (Gastropoda: Opisthobranchia) from tropical southern Africa. *Ann. S. Afr. Mus.* **95**: 233-267.

Gosliner, T., and Griffiths, R.

1981. Description and revision of some South African aeolidacean Nudibranchia (Mollusca: Gastropoda). *Ann. S. Afr. Mus.* **84**: 105-150.

Gould, A.

1859. Descriptions of shells collected in the North Pacific Exploring Expedition. *Proc. Boston Nat. Hist. Soc.* **7**: 40-45.

Griffiths, R.

1985. Description of a new South African arminacean and the proposed re-instatement of the genus *Atthila* Bergh (Mollusca: Opisthobranchia). *Ann. S. Afr. Mus.* **95**: 269-280.

Harris, L.

1968. Notes on the biology and distribution of the aeolid nudibranch (Gastropoda), *Phestilla melanobrachia* Bergh, 1874. *Publ. Seto Mar. Biol. Lab.* **16**: 193-198.

Kay, E.
1968. A review of the bivalved gastropods and a discussion of evolution within the Sacoglossa. *Symp. Zool. Soc. London* **22**: 109-134.

Kay, E. and Young, D.
1969. The Doridacea (Opisthobranchia: Mollusca) of the Hawaiian Islands. *Pac. Sci.* **23**: 172-231.

Kilburn, R., and Rippey, E.
1982. *Sea shells of southern Africa.* MacMillan, Johannesburg.

Krauss, F.
1848. *Die Südafrikanischen Mollusken. Ein Beiträg zur Kenntniss des Kap-und Natallandes und zur geographischen Verbreitung derselben, mit Beschreibung und Abbildung der neuen Arten.* Stuttgart: Ebner and Seubert.

Linnaeus, C.
1767. *Systema Naturae.* Editio duodecima reformata. "Vermes Testacea", **1**(2): 533-1327, [26]: 1106-1269.

Macnae, W.
1954*a*. On some eolidacean nudibranchiate molluscs from South Africa. *Ann. Natal Mus.* **13**: 1-50.
1954*b*. On four sacoglossan molluscs new to South Africa. *Ann. Natal Mus.* **13**: 51-64.
1955. On four species of the genus *Aplysia* common in South Africa. *Ann. Natal. Mus.* **13**: 223-241.
1957. On the occurrence of *Aplysia dactylomela* and *Aplysia oculifera* in South Africa. *S. Afr. J. Sci.* **53**: 289-292.
1958. The families Polyceridae and Goniodorididae (Mollusca, Nudibranchiata) in southern Africa. *Trans. Roy. Soc. S. Afr.* **35**: 341-372.
1962*a*. Notaspidean opisthobranchiate molluscs from South Africa. *Ann. Natal Mus.* **15**: 167-181.
1962*b*. Tectibranch molluscs from southern Africa. *Ann. Natal Mus.* **15**: 183-199.

Macnae, W., and Kalk, M.
1958. *A natural history of Inhaca Island, Moçambique.* Johannesburg: University of the Witwatersrand.

Marcus, Er.
1957. On Opisthobranchia from Brazil. 2. *J. Linn. Soc. London* **43**: 390-486.

Marcus, Ev.
1972. On some opisthobranchs from Florida. *Bull mar. Sci.* **22**: 284-308.
1976. On *Kentrodoris* and *Jorunna* (Gastropoda Opisthobranchia), *Bolm. Zool.* **1**: 11-67.
1980. Review of the western Atlantic Elysiidae (Opisthobranchia Ascoglossa), with a description of a new *Elysia* species. *Bull. Mar. Sci.* **30**: 54-79.
1983. The western Atlantic Tritoniidae. *Bolm. Zool.* **6**: 177-214.

Marcus, Ev., and Gosliner, T.
1984. Review of the family Pleurobranchaeidae (Mollusca: Opisthobranchia). *Ann. S. Afr. Mus.* **93**: 1-52.

Martens, E.
1879. Übersicht der von ihm (W. Peters) von 1843 bis 1847 in Mossambique gesammelten Mollusca. *Monat. K. Akad. Wissen.* **1879**: 727-749.

Meisenheimer, J.
1905. Pteropoda. *Wiss. Ergelon. Tief-see Exped.* **9**(1): 1-314.

Millen, S., and Gosliner, T.
1985. Four new species of dorid nudibranchs belonging to the genus *Aldisa* (Mollusca: Opisthobranchia), with a revision of the genus. *Zool. J. Linn. Soc.* **84**: 195-233.

O'Donoghue, C.
1929. Opisthobranchiate Mollusca collected by the South African Marine Biological Survey. *Union of South Africa Fisheries and Marine Biological Survey Report* **1929**: 1-84.

Pelseneer, P.
1888. Report on the Pteropoda collected by H.M.S. Challenger, during the years 1873-1876. II. The Thecosomata. *Rep. Scient. Results Voy. Challenger* **23**(65): 1-132.

Pruvot-Fol, A.
1954. Mollusques opisthobranches. *Faune Fr.* **58**: 1-460.

Quoy, J., and Gaimard, J.
1824. [In] Freycinet, Louis de. *Voyage autor du monde executé sur les corvettes de S.M. l'Uranie et la Physicienne pendant les annees 1817-1820.* Zool. pt. 2, ch. 11, Description des Mollusques: 410-516.

Rang, P.
1829. *Manuel de l'histoire naturelle des mollusques et de leurs coquilles*: Paris: 124-137.

Rao, K.P.
1965. *Moridella brockii* Bergh, 1888, redescribed with notes on anatomy and early development. *J. Mar. Biol. Assoc. India* **7**: 61-68.

Rudman, W.
1972. A comparative study of the genus *Philinopsis* Pease, 1860. (Aglajidae:Opisthobranchia) *Pacif. Sci.* **26**: 381-399.
1973. Chromodorid opisthobranch Mollusca from the Indo-West Pacific. *Zool. J. Linn. Soc.* **52**: 175-199.
1977. Chromodorid opisthobranch Mollusca from East Africa and the tropical West Pacific. *Zool. J. Linn. Soc.* **61**: 351-397.
1978. The dorid opisthobranch genera *Halgerda* Bergh and *Sclerodoris* Eliot from the Indo-West Pacific. *Zool. J. Linn. Soc.* **68**: 139-172.
1980. Aeolid opisthobranch molluscs (Glaucidae) from the Indian Ocean and South-West Pacific. *Zool. J. Linn. Soc.* **68**: 139-172.
1981*a*. Further studies on the anatomy and ecology of opisthobranch molluscs feeding on the scleractinian coral *Porites*. *Zool. J. Linn. Soc.* **71**: 373-412.
1981*b*. The anatomy and biology of alcyonarian-feeding aeolid opisthobranch molluscs and their development of symbiosis with zooxanthellae. *Zool. J. Linn. Soc.* **72**: 219-262.

Smith, E.
1902. Descriptions of new species of marine shells from South Africa. *J. Conch. London* **10**: 248-319.
1903. A list of species of Mollusca from South Africa, forming an appendix to G.B. Sowerby's 'Marine Shells of South Africa'. *Proc. Malac. Soc. London* **5:** 354-402.
1910. On South African marine Mollusca, with descriptions of new species. *Ann. Natal Mus.* **2**: 175-219.

Sowerby, G.B.
1873. Descriptions of twelve new species of shells. *Proc. Zool. Soc. London* **1873**: 718-722.
1892. *Marine shells of South Africa.* London: Sowerby.
1894. Marine shells of South Africa. *J. Conch. London* **7:** 368-378.
1897. *Appendix to Marine shells of South Africa.* London: Sowerby.

Stimpson, W.
1854. Descriptions of some new marine *Invertebrata. Proc. Acad. Nat. Sci. Philadelphia* **7:** 385-395.

Tardy, J.
1969. Un nouveau genre de nudibranche méconnu des côtes atlantique et de la Manche: *Pruvotfolia* (nov. g.) *pselliotes*, (Labbé), 1923. *Vie Milieu* **20:** 327-346.

Tesch, J.
1903. Vorläufige Mitteilung über die Thecosomata and Gymnosomata der Siboga-Expedition. *Tijdschr. Nederland. Dierk. Ver.* **8**: 111-117.

Thiele, J.
1912. Die antarktischen Schnecken und Muscheln. [in] Drygalski: *Deutsche Sudpolar-Expedition 1901-1903, Zool.* **13**(5): 183-285.
1925. Gastropoda der Deutschen Tiefsee-Expedition. Pt. 2. *Wiss. Ergebn, dt. Tiefsee-Exped.* **17**: 37-382.

Thompson, T.E.
1972. Observations on *Hexabranchus* from the Australian Great Barrier Reef. *Veliger* **15**: 1-5.
1975. Dorid nudibranchs from eastern Australia (Gastropoda, Opisthobranchia). *J. Zool. London* **176**: 477-517.
1979. Biology and relationships of the South African sacoglossan mollusc *Volvatella laguncula. J. Zool. London* **189**: 339-347.

Thompson, T.E., and Brown, G.
1981. Biology and relationships of the nudibranch mollusc *Notobryon wardi* in South Africa, with a review of the scyllaeidae. *J. Zool. London* **194**: 437-444.

Thompson, T.E., and McFarlane, I.
1967. Observations on a collection of *Glaucus* from the Gulf of Aden with a critical review of published records of *Glaucidae (Gastropoda, Opisthobranchia) Proc. Linn. Soc. London* **178**: 107-123.

Tomlin, J.
1920. Four new marine species from South Africa. *J. Conch. London* **16**: 87-88.

Turton, W.
1932. *The marine shells of Port Alfred, S. Africa.* Oxford: University Press.

Vayssière, A.
1900. Monographie de la famille des Pleurobranchidès, *2. Ann. Sci. Nat. Zool.* **12**: 1-85.

Watson, R.
1886. Report on the Scaphopoda and Gasteropoda collected by H.M.S. *Challenger* during the years 1873-1876. *Challenger Rep.* **15**: 1-756.

Willan, R.
1976. The opisthobranch *Thecacera pennigera* (Montagu) in New Zealand, with a discussion of the genus *Veliger* **18**: 347-352.

Willan, R., and Coleman, N.
1984. Nudibranchs of Australasia. Neville Coleman, Sydney.

Index

A

Acanthodoris sp., 91
Aegires sp., 99
Aeolidiella alba, 128
Aeolidiella indica, 128
Akera soluta, 46
Aldisa benguelae, 73
Aldisa trimaculata, 72
Amanda armata, 125
Ancula sp., 94
Anisodoris sp. 1, 65
Anisodoris sp. 2, 65
Aphelodoris brunnea, 66
Aphelodoris sp., 67
Aplysia dactylomela, 47
Aplysia juliana, 47
Aplysia oculifera, 47
Aplysia parvula, 46
Aplysiopsis sinusmensalis, 57
Armina gilchristi, 109
Artachaea sp., 68
Ascobulla fischeri, 49
Atagema gibba, 63
Atagema rugosa, 63

B

Baeolidia palythoae, 129
Berghia chaka, 128
Berthelinia schlumbergeri, 50
Berthella plumula, 60
Berthella sp., 60
Berthella tupala, 59
Berthellina citrina, 60
Biogeography, 13
Bonisa nakaza, 112
Bornella adamsi, 104
Bornella anguilla, 105
Bulla ampulla, 45
Bursatella leachi africana, 48
Bursatella leachi leachi, 48

C

Cadlina sp. 1, 85
Cadlina sp. 2, 86
Cadlina sp. 3, 86
Cadlina sp. 4, 86
Caloria indica, 122
Caloria sp. 1, 122
Caloria sp. 2, 123
Caloria sp. 3, 123
caryophyllidia, 21, 26
Catriona casha, 119
Catriona columbiana, 119
Catriona sp., 120
cerata, 21, 28
Ceratophyllidia africana, 91
Ceratosoma sp., 85
Chelidonura fulvipunctata, 42
Chelidonura hirundinina, 42
Chelidonura sp., 43
Chromodoris africana, 74
Chromodoris alderi, 75
Chromodoris annulata, 74
Chromodoris geometrica, 76
Chromodoris hamiltoni, 74
Chromodoris inopinata, 75
Chromodoris marginata, 76
Chromodoris sp. 1, 76
Chromodoris sp. 2, 77
Chromodoris sp. 3, 77
Chromodoris sp. 4, 77
Chromodoris sp. 5, 78
Chromodoris sp. 6, 78
Chromodoris sp. 7, 78
Chromodoris sp. 8, 79
Chromodoris vicina, 75
collection of opisthobranchs, 12
coloration, 7
Corambe sp., 95
Coryphellina sp., 114
Cratena capensis, 125
Cratena simba, 126
Cratena sp. 1, 126
Cratena sp. 2, 126
Cratena sp. 3, 127
Cratena sp. 4, 127
Creseis acicula, 46
Crimora sp. 4, 95
Cuthona annulata, 117
Cuthona kanga, 116
Cuthona ornata, 116
Cuthona sp. 1, 117
Cuthona sp. 2, 117
Cuthona sp. 3, 118
Cuthona sp. 4, 118
Cuthona sp. 5, 118
Cuthona speciosa, 116
Cylichna tubulosa, 40

D

defense mechanisms, 7
Dendrodoris caesia, 88
Dendrodoris denisoni, 88
Dendrodoris nigra, 87
Dendrodoris sp. 1, 88
Dendrodoris sp. 2, 89
Dermatobranchus sp. 1, 110
Dermatobranchus sp. 2, 110
Dermatobranchus sp. 3, 110
Dermatobranchus sp. 4, 111
detorsion, 7
Diaphorodoris sp., 92
Discodoris fragilis, 64
Discodoris sp. 1, 64
Discodoris sp. 2, 64
Dolabella auricularia, 48
Doriopsilla miniata, 89
Doriopsilla sp. 1, 89
Doriopsilla sp. 2, 90
Doriopsis pecten, 63
Doris sp. 1, 62
Doris sp. 2, 62
Doris verrucosa, 62
Doto coronata, 106
Doto pinnatifida, 106
Doto rosea, 107
Durvilledoris lemniscata, 79

E

egg masses, 9, 10
Elysia halimedae, 52
Elysia livida, 53
Elysia marginata, 52
Elysia moebii, 53
Elysia rufescens, 54
Elysia sp. 1, 54
Elysia sp. 2, 54
Elysia sp. 3, 55
Elysia sp. 4, 55
Elysia sp. 5, 55
Elysia sp. 6, 56
Elysia sp. 7, 56
Elysia vatae, 53
Elysia virgata, 52
Elysia viridis, 51
Embletonia gracilis, 115
Eubranchus sp. 1, 114
Eubranchus sp. 2, 115
Eubranchus sp. 3, 115
evolution, 7

F

Facelina olivacea, 121
Facelina sp., 121
Favorinus ghanensis, 124
Favorinus japonicus, 123
Fiona pinnata, 120
Flabellina capensis, 113
Flabellina funeka, 112
Flabellina sp. 1, 113
Flabellina sp. 2, 113
Flabellina sp. 3, 114

G

Gargamella sp. 1, 70
Gargamella sp. 2, 71
Gastropteron alboaurantium, 43
Gastropteron flavobrunneum, 43
Geitodoris capensis, 65
Glaucus atlanticus, 127
Glossodoris atromarginata, 83
Glossodoris sp. 1, 84
Glossodoris sp. 2, 84
Glossodoris sp. 3, 84
Glossodoris sp. 4, 85
Godiva quadricolor, 124
Goniodoris castanea, 93
Goniodoris mercurialis, 93
Goniodoris sp., 93
Gymnodoris alba, 102
Gymnodoris inornata, 102
Gymnodoris okinawae, 102
Gymnodoris sp. 1, 103
Gymnodoris sp. 2, 103

H

Halgerda formosa, 69
Halgerda punctata, 69
Halgerda wasinensis, 68
Hallaxa sp., 69
Haminoea alfredensis, 44
Haminoea natalensis, 44
Hexabranchus sanguineus, 87
Hydatina amplustre, 40
Hydatina physis, 40
Hypselodoris capensis, 81
Hypselodoris carnea, 81
Hypselodoris infucata, 81
Hypselodoris maridadilus, 82
Hypselodoris sp. 1, 82
Hypselodoris sp. 2, 82
Hypselodoris sp. 3, 83

I

introduced species, 13

J

Janolus capensis, 111
Janolus longidentatus, 112
Japonacteon sp., 39
Jorunna tomentosa, 70
Jorunna zania, 70
Julia zebra, 50

K

Kaloplocamus ramosus, 98

L

Lecithophorus capensis, 103
Lecithophorus sp., 104
Leminda millecra, 111
Limacia clavigera, 99
Lobiger souverbiei, 50

M

Marianina rosea, 107
Marionia sp. 1, 108
Marionia sp. 2, 109
Marioniopsis cyanobranchiata, 109
Melibe rosea, 105
Melibe sp., 106
Metaruncina sp., 45
Micromelo undata, 39
Moridilla brockii, 122
Mourgona sp., 57

N

Nembrotha livingstonei, 99
Nembrotha purpureolineata, 100
Notobryon wardi, 105
Noumea purpurea, 80
Noumea sp., 80
Noumea varians, 79

O

Okadaia elegans, 104
Okenia amoenula, 92
Okenia sp., 92
Oxynoe sp., 51
Oxynoe viridis, 51

P

Phaneropthalmus smaragdinus, 45
Phestilla melanobrachia, 120
Philine aperta, 41
Philinopsis capensis, 42
Philinopsis cyanea, 41
Phyllidia sp. 1, 90
Phyllidia sp. 2, 91
Phyllidia varicosa, 90
Phyllobranchillus orientalis, 57
Phyllodesmium hyalinum, 125
Phyllodesmium serratum, 124
Placida dendritica, 56
Platydoris cruenta, 73
Platydoris sp., 73
Pleurobranchaea bubala, 61
Pleurobranchaea tarda, 61
Pleurobranchus inhacae, 58
Pleurobranchus nigropunctatus, 58
Pleurobranchus peroni, 59
Pleurobranchus xhosa, 59
Plocamophorus maculatus, 98
Plocamophorus sp., 98
Polycera capensis, 96
Polycera hedgpethi, 97
Polycera quadrilineata, 97
Polycera sp., 97
prey, 9
Pruvotfolia pselloites, 121

R

radula, 8, 9, 21
reproduction, 9
Retusa truncatula, 41
rhinophores,21, 24
Rictaxis albus, 39
Risbecia pulchella, 83
Roboastra gracilis, 101
Roboastra luteolineata, 101
Rostanga muscula, 71
Rostanga sp. 1, 71
Rostanga sp. 2, 72
Rostanga sp. 3, 72

S

Sclerodoris apiculata, 67
Sclerodoris coriacea, 67
Sclerodoris sp., 68
shell, 7, 8, 22
Smaragdinella calyculata, 44
sperm groove, 21, 23
Stylocheilus longicauda, 49
swimming behavior, 7

T

Tambja capensis, 100
Tambja morosa, 100
Tambja sp., 101
Tergipes tergipes, 119
Thecacera pacifica, 96
Thecacera pennigera, 95
Thecacera sp., 96
Thordisa sp. 1, 66
Thordisa sp. 2, 66
Thorunna horologica, 80
Trapania sp. 1, 94
Trapania sp. 2, 94
Tritonia nilsodhneri, 107
Tritonia sp. 1, 108
Tritonia sp. 2, 108

U

Umbraculum sinicum, 58

V

veliger larva, 11
Volvatella laguncula, 49

Map. Collecting localities in southern Africa.

1. Luderitz
2. Port Nolloth
3. Elands Bay
4. Saldanha Bay/Langebaan Lagoon
5. Cape Town
6. Bloubergstrand
7. Milnerton
8. Bakoven
9. Llandudno
10. Kommetjie
11. Cape Point
12. Buffles Bay
13. Venus Pool
14. Castle Rocks
15. Miller's Point
16. Dassie Point
17. Simonstown
18. Clovelly
19. Dale Brook
20. St. James
21. Muizenberg
22. Strandfontein
23. Rooi Els
24. Cape Hangklip
25. Hermanus
26. Knysna
27. Plettenberg Bay
28. Jeffreys Bay
29. Port Elizabeth/Algoa Bay
30. Port Alfred
31. East London
32. Gonubie
33. Coffee Bay
34. Umtata River Mouth
35. Umgazana
36. Port St. Johns
37. Scottburgh/Aliwal Shoals
38. Park Rynie
39. Durban
40. Salt Rock/Chaka's Rock
41. Sodwana Bay National Park
42. Kosi Bay

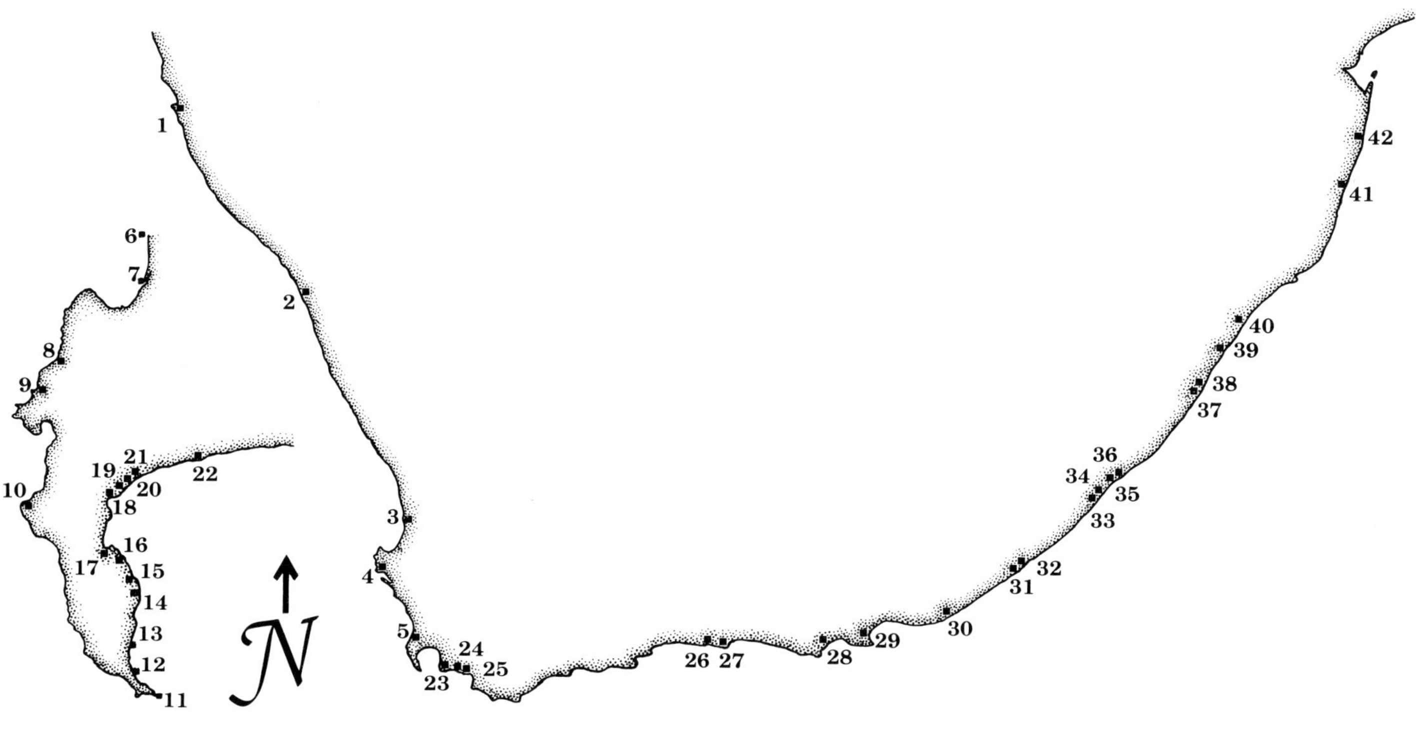

Sizes of Animals Illustrated

1. Rictaxis albus, 22 mm
2. Japonacteon sp., 11mm
3. Micromelo undata, 13 mm
4. Hydatina physis, 35mm
5. Hydatina amplustre, 40 mm
6. Cylichna tubulosa, 8 mm
7. Retusa truncatula, 3 mm
8. Philine aperta, 50 mm
9. Philinopsis cyanea, 45 mm
10. Philinopsis capensis, 11 mm
11. Chelidonura hirundinina, 20 mm
12. Chelidonura fulvipunctata, 25 mm
13. Chelidonura sp., 15 mm
14. Gastropteron flavobrunneum, 7 mm
15. Gastropteron alboaurantium, 5 mm
16. Haminoea alfredensis, 15 mm
17. Haminoea natalensis, 12 mm
18. Smaragdinella calyculata, 8 mm
19. Phaneropthalmus smaragdinus, 18 mm
20. Metaruncina sp., 3 mm
21. Bulla ampulla, 14 mm
22. Creseis acicula, 20 mm
23. Akera soluta, 60 mm
24. Aplysia parvula, 35 mm
25. Aplysia oculifera, 100 mm
26. Aplysia dactylomela, 115 mm
27. Aplysia juliana, 60 mm
28. Dolabella auricularia, 150 mm
29. Bursatella leachi, 118 mm, 130 mm
30. Stylocheilus longicauda, 28 mm
31. Ascobulla fischeri, 8 mm
32. Volvatella laguncula, 15 mm
33. Berthelinia schlumbergeri, 26 mm
34. Julia zebra, 10 mm
35. Lobiger souverbiei, 7 mm
36. Oxynoe viridis, 12 mm
37. Oxynoe sp., 35 mm
38. Elysia viridis, 7 mm
39. Elysia halimedae, 10 mm
40. Elysia marginata, 7 mm
41. Elysia virgata, 13 mm
42. Elysia moebii, 12 mm
43. Elysia vatae, 9 mm
44. Elysia livida, 17 mm
45. Elysia rufescens, 8 mm
46. Elysia sp. 1, 18 mm
47. Elysia sp. 2, 10 mm
48. Elysia sp. 3, 9 mm
49. Elysia sp. 4, 8 mm
50. Elysia sp. 5, 5 mm
51. Elysia sp. 6, 6 mm
52. Elysia sp. 7, 18 mm
53. Placida dendritica, 6 mm
54. Aplysiopsis sinusmensalis, 12 mm
55. Phyllobranchillus orientalis, 30 mm
56. Mourgona sp., 5 mm
57. Umbraculum sinicum, 40 mm
58. Pleurobranchus inhacae, 25 mm
59. Pleurobranchus nigropuntatus, 40 mm
60. Pleurobranchus peroni, 50 mm
61. Pleurobranchus xhosa, 32 mm
62. Berthella tupala, 13 mm
63. Berthella plumula, 20 mm
64. Berthella sp., 17 mm
65. Berthellina citrina, 32 mm
66. Pleurobranchaea tarda, 25 mm
67. Pleurobranchaea bubala, 60 mm
68. Doris verrucosa, 32 mm
69. Doris sp. 1, 4 mm
70. Doris sp. 2, 25 mm
71. Doriopsis pecten, 12 mm
72. Atagema rugosa, 12 mm
73. Atagema gibba, 40 mm
74. Discodoris fragilis, 60 mm
75. Discodoris sp. 1, 28 mm
76. Discodoris sp. 2, 22 mm
77. Geitodoris capensis, 26 mm
78. Anisodoris sp. 1, 40 mm
79. Anisodoris sp. 2, 16 mm
80. Thordisa sp. 1, 12 mm
81. Thordisa sp. 2, 42 mm
82. Aphelodoris brunnea, 45 mm
83. Aphelodoris sp., 40 mm
84. Sclerodoris apiculata, 55 mm
85. Sclerodoris coriacea, 38 mm
86. Sclerodoris sp., 30 mm
87. Artachaea sp., 48 mm
88. Halgerda wasinensis, 45 mm
89. Halgerda formosa, 19 mm
90. Halgerda punctata, 18 mm
91. Hallaxa sp., 6 mm
92. Jorunna zania, 62 mm
93. Jorunna tomentosa, 35 mm
94. Gargamella sp. 1, 20 mm
95. Gargamella sp. 2, 18 mm
96. Rostanga muscula, 11 mm
97. Rostanga sp. 1, 20 mm
98. Rostanga sp. 2, 40 mm
99. Rostanga sp. 3, 25 mm
100. Aldisa trimaculata, 18 mm
101. Aldisa benguelae, 17 mm
102. Platydoris cruenta, 40 mm
103. Platydoris sp., 30 mm
104. Chromodoris hamiltoni, 28 mm
105. Chromodoris africana, 45 mm
106. Chromodoris annulata, 30 mm
107. Chromodoris inopinata, 20 mm
108. Chromodoris alderi, 17 mm
109. Chromodoris vicina, 23 mm
110. Chromodoris marginata, 17 mm
111. Chromodoris geometrica, 16 mm
112. Chromodoris sp. 1, 25 mm
113. Chromodoris sp. 2, 13 mm
114. Chromodoris sp. 3, 20 mm
115. Chromodoris sp. 4, 14 mm
116. Chromodoris sp. 5, 7 mm
117. Chromodoris sp. 6, 23 mm
118. Chromodoris sp. 7, 30 mm
119. Chromodoris sp. 8, 6 mm
120. Durvilledoris lemniscata, 23 mm
121. Noumea varians, 18 mm
122. Noumea purpurea, 5 mm
123. Noumea sp., 12 mm
124. Thorunna horologica, 17 mm
125. Hypselodoris capensis, 40 mm
126. Hypselodoris carnea, 23 mm
127. Hypselodoris infucata, 35 mm
128. Hypselodoris maridadilus, 18 mm
129. Hypselodoris sp. 1, 26 mm
130. Hypselodoris sp. 2, 32 mm
131. Hypselodoris sp. 3, 27 mm
132. Risbecia pulchella, 52 mm
133. Glossodoris atromarginata, 11 mm
134. Glossodoris sp. 1, 30 mm
135. Glossodoris sp. 2, 27 mm
136. Glossodoris sp. 3, 42 mm
137. Glossodoris sp. 4, 18 mm
138. Ceratosoma sp., 26 mm
139. Cadlina sp. 1, 20 mm
140. Cadlina sp. 2, 17 mm
141. Cadlina sp. 3, 24 mm
142. Cadlina sp. 4, 52 mm
143. Hexabranchus sanguineus, 78 mm
144. Dendrodoris nigra, 65 mm
145. Dendrodoris caesia, 70 mm
146. Dendrodoris denisoni, 42 mm
147. Dendrodoris sp. 1, 15 mm
148. Dendrodoris sp. 2, 7 mm
149. Doriopsilla miniata, 28 mm
150. Doriopsilla sp. 1, 52 mm
151. Doriopsilla sp. 2, 17 mm
152. Phyllidia varicosa, 62 mm
153. Phyllidia sp. 1, 35 mm
154. Phyllidia sp. 2, 42 mm
155. Ceratophyllidia africana, 41 mm
156. Acanthodoris sp., 16 mm
157. Diaphorodoris sp., 5 mm
158. Okenia amoenula, 23 mm
159. Okenia sp., 10 mm
160. Goniodoris mercurialis, 7 mm
161. Goniodoris castanea, 20 mm
162. Goniodoris sp., 3 mm
163. Trapania sp. 1, 7 mm
164. Trapania sp. 2, 10 mm
165. Ancula sp., 7 mm
166. Corambe sp., 7 mm
167. Crimora sp., 8 mm
168. Thecacera pennigera, 11 mm
169. Thecacera pacifica, 60 mm
170. Thecacera sp., 13 mm
171. Polycera capensis, 50 mm
172. Polycera quadrilineata, 8 mm
173. Polycera hedgpethi, 30 mm
174. Polycera sp., 21 mm
175. Plocamophorus maculatus, 40 mm
176. Plocamophorus sp., 13 mm
177. Kaloplocamus ramosus, 42 mm
178. Aegires sp., 7 mm
179. Limacia clavigera, 26 mm
180. Nembrotha livingstonei, 20 mm
181. Nembrotha purpureolineata, 35 mm
182. Tambja capensis, 30 mm
183. Tambja morosa, 20 mm
184. Tambja sp., 40 mm
185. Roboastra gracilis, 30 mm
186. Roboastra luteolineata, 18 mm
187. Gymnodoris inornata, 27 mm
188. Gymnodoris okinawae, 18 mm
189. Gymnodoris alba, 10 mm
190. Gymnodoris sp. 1, 10 mm
191. Gymnodoris sp. 2, 11 mm
192. Lecithophorus capensis, 13 mm
193. Lecithophorus sp., 8 mm
194. Okadaia elegans, 3 mm
195. Bornella adamsi, 24 mm
196. Bornella anguilla, 42 mm
197. Notobryon wardi, 30 mm
198. Melibe rosea, 36 mm
199. Melibe sp., 60 mm
200. Doto coronata, 6 mm
201. Doto pinnatifida, 22 mm
202. Doto rosea, 10 mm
203. Marianina rosea, 12 mm
204. Tritonia nilsodhneri, 30 mm
205. Tritonia sp. 1, 30 mm
206. Tritonia sp. 2, 30 mm
207. Marionia sp. 1, 11 mm
208. Marionia sp. 2, 15 mm
209. Marioniopsis cyanobranchiata, 60 mm
210. Armina gilchristi, 35 mm
211. Dermatobranchus sp. 1, 20 mm
212. Dermatobranchus sp. 2, 35 mm
213. Dermatobranchus sp. 3, 17 mm
214. Dermatobranchus sp. 4, 15 mm
215. Leminda millecra, 55 mm
216. Janolus capensis, 25 mm
217. Janolus longidentatus, 35 mm
218. Bonisa nakaza, 100 mm
219. Flabellina funeka, 20 mm
220. Flabellina capensis, 20 mm
221. Flabellina sp. 1, 9 mm
222. Flabellina sp. 2, 11 mm
223. Flabellina sp. 3, 7 mm
224. Coryphellina sp., 16 mm
225. Eubranchus sp. 1, 8 mm
226. Enbranchus sp. 2, 6 mm
227. Enbranchus sp. 3, 5 mm
228. Embletonia gracilis, 7 mm
229. Cuthona speciosa, 20 mm
230. Cuthona ornata, 6 mm
231. Cuthona kanga, 14 mm
232. Cuthona annulata, 6 mm
233. Cuthona sp. 1, 5 mm
234. Cuthona sp. 2, 6 mm
235. Cuthona sp. 3, 20 mm
236. Cuthona sp. 4, 4 mm
237. Cuthona sp. 5, 7 mm
238. Tergipes tergipes, 6 mm
239. Catriona casha, 8 mm
240. Catriona columbiana, 7 mm
241. Catriona sp., 7 mm
242. Phestilla melanobrachia, 11 mm
243. Fiona pinnata, 20 mm
244. Facelina olivacea, 23 mm
245. Facelina sp., 11 mm
246. Pruvotfolia pselliotes, 15 mm
247. Moridilla brockii, 35 mm
248. Caloria indica, 18 mm
249. Caloria sp. 1, 22 mm
250. Caloria sp. 2, 30 mm
251. Caloria sp. 3, 15 mm
252. Favorinus japonicus, 11 mm
253. Favorinus ghanensis, 7 mm
254. Godiva quadricolor, 20 mm
255. Phyllodesmium serratum, 35 mm
256. Phyllodesmium hyalinum, 8 mm
257. Amanda armata, 12 mm
258. Cratena capensis, 24 mm
259. Cratena simba, 8 mm
260. Cratena sp. 1, 12 mm
261. Cratena sp. 2, 7 mm
262. Cratena sp. 3, 7 mm
263. Cratena sp. 4, 20 mm
264. Glaucus atlanticus, 30 mm
265. Aeolidiella indica, 35 mm
266. Aeolidiella alba, 7 mm
267. Berghia chaka, 8 mm
268. Baeohdia palythoae, 13 mm